Asphalt Mixture Specification and Testing

Asphalt Mixture Specification and Testing

J. Cliff Nicholls

CRC Press
Taylor & Francis Group
Boca Raton London New York

CRC Press is an imprint of the
Taylor & Francis Group, an **informa** business

CRC Press
Taylor & Francis Group
6000 Broken Sound Parkway NW, Suite 300
Boca Raton, FL 33487-2742

First issued in paperback 2019

ISBN-13: 978-0-4987-6405-6 (hbk)
ISBN-13: 978-0-367-87763-7 (pbk)

Library of Congress Cataloging-in-Publication Data

Names: Nicholls, Cliff, author.
Title: Asphalt mixture specification and testing / Cliff Nicholls.
Description: Boca Raton : CRC Press, 2017. | Includes bibliographical references and index.
Identifiers: LCCN 2016047387 | ISBN 9781498764056 (hardback : acid-free paper)
Subjects: LCSH: Pavements, Asphalt--Testing. | Pavements, Asphalt--Specifications.
Classification: LCC TE270 .N47 2017 | DDC 625.8/50212--dc23
LC record available at https://lccn.loc.gov/2016047387

Visit the Taylor & Francis Web site at
http://www.taylorandfrancis.com

and the CRC Press Web site at
http://www.crcpress.com

This book is dedicated to my wife, Carol, not so much for the support and encouragement that she has given me in writing it (her encouragement was definitely forthcoming) but more to placate her for not accepting her suggestion of 'Fifty shades of asphalt' as the title.

I am very grateful to John Prime for generously providing me with photographs of test equipment and to Pippa Birch, Robert Hunter, Thanos Nikolaides and Ian Walsh in proof reading early versions of the book and offering useful suggestions to improve it. All of them are true friends.

I also thank my fellow members, past and present, of CEN TC227/WG1/TG2, the European task group responsible for the many parts of EN 12697, for drafting these test standards. This test series forms the background for much of this book as well as being the justification for my wife's alternative title because there are around fifty parts!

Contents

List of tables

List of figures

Preface

Asphalt is a complex material that has many beneficial structural and serviceability properties for use in road, airfield and other surfaces. Asphalt also has a good sustainability record with the ability to be completely recycled or to incorporate various secondary materials. However, some of the properties are mutual exclusive, and increasing the number of properties required increases the difficulty to produce a complying mixture and, therefore, the cost. Different pavement layers require different properties, as do different locations and categories of pavement. There are several different ways that each of these properties can be specified.

This book reviews each property in terms of the different ways that the property can be specified, from recipe and processing requirements to test methods on the asphalt or its constituents which can be surrogate, simulative or fundamental. The description of the tests that can be used includes advantages and limitations as far as measuring the desired property as well as an indication of the levels that can be set and the precision that has been found. The tests will be mainly based around the European test methods including the EN 12697 suite of asphalt tests.

The aim of the book is to encourage everybody in the construction process to be able to consider what is really needed to specify and produce asphalt that is appropriate for the intended use.

Author

Dr. J. Cliff Nicholls, known generally as Cliff, was educated at King's School Worcester and Imperial College of Science and Technology. Cliff graduated in civil engineering from Imperial College in 1972 to join Rendel Palmer & Tritton, before moving to the Property Services Agency, the Department of the Environment, the Building Research Establishment and finally the Transport and Road Research Laboratory (now TRL Limited). From 1988, he has been researching pavement materials, in particular asphalt and asphalt test methods. Cliff retired from TRL in October 2015 although he remains on a call-off contract.

Prior to retirement, Cliff was a Senior Academy Fellow in the Infrastructure Division at TRL Limited. He was mainly involved in research into asphalt surface course materials, often as project manager. Projects have involved a wide range of materials, including hot rolled asphalt to surface dressing and including porous asphalt, high-friction surfacings, thin surface course systems and stone mastic asphalt as well as some on associated materials such as road markings.

Cliff sat on the British Standards Institution and Comité Européen de Normalisation committees for asphalt, including being the convenor of the CEN test methods task group for test methods from 2000 to 2015. He was also a member of several British Board of Agrément Highway Authorities Products Approval Scheme steering groups and was a member of the Council of the Institute of Asphalt Technology. Cliff has written many TRL reports and other learned papers. He edited the book *Asphalt Surfacings* which was published by E & FN Spon in 1998 and has contributed to other books. He obtained a DPhil by published works from the University of Ulster in August 1999.

Chapter 1

Introduction

1.1 OBJECTIVE

Bitumen is a 'hydrocarbon product produced from the refining of crude oil' that is 'a thermoplastic, viscoelastic liquid that behaves as glass-like solid at low temperatures and/or during short loading times and as a viscous fluid at high temperatures and/or during short loading' (Hunter et al., 2015), making it one of the most complex construction materials. When it is combined into asphalt, which is a 'surfacing material consisting of bitumen, mineral aggregates and fillers and may contain other additives' (Hunter et al., 2015) for which the aggregates can have a variety of shapes, gradings and compositions, that asphalt becomes an even more complex material that has been, and is being, used successfully on roads, airfields and other paved areas. However, the properties asked of asphalt pavements (the number of which seems to be continually increasing with time) differ both for different layers within the pavement and for different circumstances and, therefore, require different asphalt mixtures.

The objective of this book is to review the ways that asphalt can be specified with particular emphasis on the test methods used to measure the performance of the various properties. Therefore, the tests are described in terms of their advantages and limitations as far as measuring the principal desired properties. It is hoped that this approach will help engineers define what is required of the material needed for their specific situations without requesting any properties, or levels of properties, that are not necessary for that situation.

As such, a résumé of specifications and their relative advantages and disadvantages for different situations is given. Then, different properties are discussed in terms of

- Their specification
- The test methods that can be used (primarily the EN 12697 suite of European methods)
- The extent to which the results predict performance

- The categories that can be set for the test when assessing different asphalt types
- The precision that has been found for different tests
- The other properties that are adversely affected by enhanced performance

Finally, various aspects about sustainability are discussed, with a strong emphasis on durability. It is hoped that better understanding of the sustainability needs will encourage improved and economical implementation of the means to produce durable and sustainable asphalt pavements.

1.2 TERMINOLOGY AND UNITS

Oscar Wilde described the United Kingdom and United States as 'two nations divided by a common language' and nowhere is this statement more relevant than for asphalt technology. The two countries use different terms to mean the same thing for several items, and these differences have been further exacerbated by the harmonisation of terms across Europe requiring the United Kingdom to change some terms that were previously the same as those used in the United States. Therefore, it is important to clarify which terms are being adopted in any publication.

The main differences are with regard to the materials and the layers in which that material is used. The term 'asphalt' is used in Europe solely to mean the mixture including the aggregate whereas in America it can be used to mean either the mixture, when it is generally introduced as 'asphalt concrete', 'asphaltic concrete' or 'hot mix asphalt (HMA)' or the binder, when it is generally introduced as 'asphalt cement'. In Europe, 'asphalt concrete' is a specific mixture type (Section 2.6.2) which can cause confusion when it is uncertain whether the term relates to the mixture type or to all/any of them, while the binder is known as 'bitumen'.

With regard to layers, the European terminology is 'surface course' at the top, 'binder course' next and 'base' (which may be split into upper base and lower base), whereas the American terms, which were previously used in the United Kingdom, are 'wearing course', 'basecourse' and 'roadbase'. There is potential for confusion if the term 'base course' is used as to whether it is the second layer with the space omitted or the bottom layer with 'course' erroneously added.

The terms are listed in Table 1.1 for quick reference. For this book, the European terms will generally be used unless stated otherwise.

Another term for the mixture, used in both America and Europe, is 'HMA'. However, the term has become more specific with the recent rise of asphalt mixtures that are mixed and laid at temperatures lower than those that were previously standard. These new reduced temperature mixtures are generally categorised as warm mix asphalt if still mixed above 100°C,

Table 1.1 Conflicts in terminology

Description	European		American
Mixture	Asphalt		Asphalt(ic) concrete
Recycled mixture	Reclaimed asphalt (RA)		Reclaimed asphalt pavement (RAP)
Binder	Bitumen		Asphalt cement
Top layer	Surface course	} Surfacing	Wearing course
Second layer	Binder course		Basecourse
Bottom layer(s)	Base		Roadbase

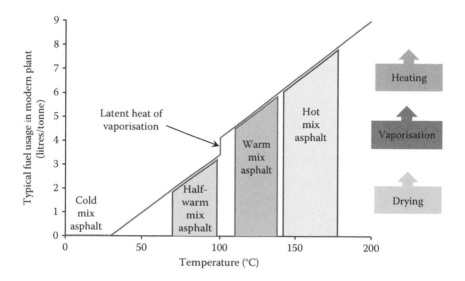

Figure 1.1 Categories of reduced temperature asphalt.

half-warm mix asphalt if mixed below 100°C and cold mix asphalt if mixed around ambient temperatures, as shown in Figure 1.1. However, the temperature at which the asphalt is mixed does not affect the properties required of them in service, so these terms will not be used in this book.

With regard to units, this book has been written with all units converted to SI units (which excludes centimetres, despite the unit sometimes being mistakenly used in standards) whenever possible to keep the units consistent. Therefore, where American tests are referenced, it is the metric conversion that has been quoted rather than the imperial units.

REFERENCE

Hunter, R N, A Self and J Read. 2015. *The Shell Bitumen Handbook*. 6th edition. London: ICE Publishing.

Chapter 2

Specifications for asphalt

2.1 DEVELOPMENT OF SPECIFICATIONS

Asphalt specifications are designed to define an asphalt mixture with the capability to both perform for all required properties and not to fail in any of the potential failure mechanisms. Early asphalt specifications were based on attempts to replicate mixtures that had already proven themselves in practice. This approach implies specifying the same constituent materials that are mixed in the same proportions using the same procedures for mixing, transportation, laying and compaction under the same temperature and climatic conditions. This approach is generally called a recipe or recipe-type specification and puts all the responsibility for performance on the mixture designer/specifier provided the contractor has followed the specification precisely. This type of specification is referred to as 'conventional' in this book because they refer to a recipe for asphalt, an aspect of the recipe and/or a methodology for producing or laying the asphalt.

Conventional specifications can work well in areas where available constituent materials, in particular aggregates, are sufficiently similar not to affect the properties of the asphalt. Once the area for which the specification is to cover includes different component materials, some of the requirements, for example the grading envelope, have to be widened in order to maintain the final asphalt properties. However, the specification becomes a compromise between widening the specification sufficiently to ensure all successful mixtures can be included and restricting the specification to exclude all unsuccessful mixtures.

Furthermore, conventional specifications are ideal in a conservative society with no innovation in which the conditions, including traffic loading or climate, do not change. When conditions change and/or when it is intended to incorporate alternative component materials, it is not practical to lay such materials, possibly in trials, and wait to monitor for the desired service life before they can be incorporated into the specification, by which time the conditions and the available materials may have changed again.

In order to allow the specification to apply more widely and maximise the use of materials available locally, performance requirements have to be added. The performance is generally measured by fundamental, simulative or surrogate tests for different aspects of the performance, moving some of the responsibility for performance to the supplier. The differences between these types of measure, including conventional requirements, are

- *Fundamental.* Tests that measure the property of interest directly (e.g. stiffness modulus for use in the design of pavements).
- *Simulative.* Tests that imitate the failure mechanism associated with the property of interest in a controlled manner (e.g. wheel-tracking rate for the property of deformation resistance).
- *Surrogate.* Tests that measure a different property to the property of interest but which is regarded as related to that property and can be measured more easily (e.g. air voids content as a measure for the property of durability and other properties).
- *Conventional.* Compositional and/or application requirements that will result in the mixture laid being as similar as possible to previously successfully laid asphalt without any direct measure of properties (e.g. bitumen grade and/or roller pattern for a number of properties).

However, the addition of performance requirements to a conventional specification can result in 'double jeopardy' in that the supplier has to follow the recipe and the pre-defined recipe has to meet the performance requirements. Therefore, the conventional requirements have to be relaxed when performance tests are added, although it is often difficult to know which aspect of those conventional requirements can be relaxed because the performance requirement will cover it. Different aspects of the menu are not specifically related to different performance aspects of the asphalt.

Another aspect is whether to measure the properties in the laboratory in order to avoid having to remove unsatisfactory material or to measure them in the as-laid mat. Removing unsatisfactory material after it has been laid is both expensive and time consuming. Measuring the properties in the as-laid mat ensures that those requirements are actually provided, including the workmanship on site. There is also the halfway house of testing trial areas before the main works are undertaken, although checks may then be needed to ensure the same workmanship is provided on the main works as the trial area. The choice can be dictated by the test required to measure the property only being possible on site, but most laboratory tests can be undertaken on

- Asphalt manufactured and compacted in the laboratory as test samples
- Plant-manufactured asphalt compacted in the laboratory as test samples
- Plant-manufactured asphalt compacted on site from which samples are cored or cut

- Plant-manufactured asphalt compacted on site from which material is taken and re-compacted in the laboratory as test samples

The problem with performance requirements is obtaining test results which accurately reflect the *in situ* performance of the asphalt in relation to the required property. In order to do that, the test must be directly applicable and be discriminatory with due allowance for the test precision. In general, the available tests rarely achieve that goal.

The ideal is that simulative and surrogate test requirements should be replaced by fundamental test requirements that measure the inherent properties themselves. This aim is obviously desirable, but is not as simple as it may appear because first the fundamental properties must be defined. Two examples that demonstrate the difficulties are air voids content and fatigue resistance. For air voids content, the voids that are to be included or excluded need to be decided: occluded voids within the aggregate particles, occluded voids within the mastic, interconnecting voids within the mastic and/or voids within the surface texture. Different methods of measuring the density, and hence air voids content, will include different combinations, with only the interconnecting voids always being included. For fatigue resistance, which is regularly included when fundamental properties are discussed, the ranking of mixtures will depend on both which action is being applied (bending, torsion, etc.) and whether the test conditions are controlled stress or controlled strain. In both these cases, the fundamental property needs greater definition than is usually the case.

In the first published version of the European standard for asphalt concrete, EN 13108-1 (CEN, 2006), the mixture specification was divided into fundamental and empirical options, with the intention that engineers could either define the mixture by fundamental or simulative/surrogate methods but not a combination of the two. This approach is flawed in that it is difficult to define the fundamental property for some required properties (e.g. fuel resistance) and the standing of the fundamental test for some performance requirements are more widely accepted than others. It is better to specify using the best available measure for each requirement, whether fundamental, simulative or conventional, than to try to stick to one approach for everything. The first revision of EN 13108-1 (CEN, 2016a) did not keep this distinction between solely fundamental and empirical options.

The ideal of measuring the potential of an asphalt mixture to perform for all required properties and not to fail in any of the potential failure mechanisms requires knowledge of all the required properties and potential failure mechanisms of the asphalt, not just when installed but after ageing. If all these properties and mechanisms were known with accurate measurements of each, there would be no need for any conventional components within a specification and the resulting specification would be blind to the mixture type (Section 2.6), the type being defined by the performance level set for each property (Section 2.2).

When performance-based specifications, whether using surrogate, simulative or fundamental tests, are used to define the design of a mixture, compliance checks for the mixture actually delivered to site are generally checks on the conventional requirements because such tests are quicker and cheaper than repeating the performance tests. However, these compliance checks are against the mixture previously tested for performance and not an idealised recipe from the specification.

2.2 UNCERTAINTY OF TEST RESULTS

There are several ways to produce asphalt specimens for testing, and the method used has an effect on the value reached for some performance tests (Gourdon et al., 1999). The principal alternatives for producing compacted specimens are set out in Table 2.1.

Uncompacted specimens only have the options of plant or laboratory mixing, but can still have variability from factors including the temperature of mixing and the length of time that the mixture is held at that temperature. All specimens can also be affected by length of time between manufacture and testing.

Even if the samples are mixed and compacted in the same way, the test results from otherwise similar samples can differ. Part of this variability is because asphalt is an inhomogeneous mixture of inhomogeneous natural component materials. In particular, the aggregate particles are not of a regular size and shape and the top surface of a specimen is unlikely to be smooth so the dimensions are imprecise. Another part of the variability is from the precision of the test method itself. This precision is given in most test standards as the repeatability and the reproducibility where

- Repeatability is the value less than or equal to which the absolute difference between two test results obtained with the same method on identical test items in the same laboratory by the same operator using the same equipment within short intervals of time may be expected to be within a probability of 95%.
- Reproducibility is the value less than or equal to which the absolute difference between two test results obtained with the same method on

Table 2.1 Alternative methods of sample preparation

Mixing	Compaction	Sampling
Plant mixed	Roller compacted on site	Coring or cut out slab
Plant or laboratory mixed	Impact compaction	None
	Slab compaction	None or coring
	Vibratory compaction	None

identical test items in different laboratories with different operators using different equipment may be expected to be within a probability of 95%.

The reproducibility can be a significant proportion of the test result value.

Given that the values for properties being specified are not specific values but a statistical frequency distribution, the specified value could be the mean or a characteristic value (usually the 95% characteristic value when the measured valued will not exceed the specified value more than once every 20 tests). What the specified value theoretically cannot be is the maximum (or minimum, depending on the property in question) value that will never be exceeded (or will always be exceeded).

2.3 PROPERTIES RELEVANT TO LAYERS

Asphalt is a very complex material that can be designed with a number of different attributes. As such, there are many properties that can be specified with different properties being required for different layers. The principal properties are shown in Table 2.2.

'Yes' implies that some performance is required, although it may not be explicitly specified for lower levels of the property.

These properties, together with the compliance checks to ensure the mixture delivered to site is what was designed to have the required performance, are discussed in detail in subsequent chapters.

2.4 LEVELS OF SPECIFICATION

The general specification, whether national or other, needs to be refined for the job specification. In the general specification, there may be conventional options or different classes for performance requirements for which selection is required. The tendency is for specifiers to select the best of everything in order to ensure that they get what they want. However, asking for the best of everything adds to the cost and may even make supply impossible because some properties are not compatible. Improving the fatigue resistance generally reduces the stiffness modulus and resistance to permanent deformation while noise- and spray-reduction usually require open-textured mixtures which have relatively poor stiffness moduli. When compiling a job specification, the needs of the site need to be explicitly considered so that the asphalt performs to the required level for the anticipated service life. There is no point in demanding resistance to the scuffing from an articulated lorry in a residential cul-de-sac or the highest level of noise reduction in a remote location or in the middle of a heavy industrial area.

Table 2.2 Properties required by pavement layers

Property	Surface course	Binder course	Base
Level	Yes	Yes	Yes
Roughness	Yes	No	No
Skid resistance	Yes	Temporarily[a]	No
Noise reduction	If required	No	No
Spray reduction	If required	No	No
Colour	If required	No	No
Deformation resistance	Yes	Yes	No[b]
Tensile strength	If required	If required	If required
Stiffness modulus	Yes[c]	Yes	Yes
Fatigue resistance	If required	Yes	Yes
Resistance to low temperature cracking	Yes	Yes	Yes
Resistance to crack development	Yes	Yes	Yes
Resistance to reflective cracking	Yes	Yes	Yes
Moisture sensitivity	Yes	Yes	Yes
Aggregate-binder affinity	Yes	Yes	Yes
Impermeability	Yes	Yes	Yes
Resistance to ravelling and particle loss	Yes	No	No
Resistance to abrasion by studded tyres	Yes[d]	No	No
Resistance to fuel and deicing fluids	Yes	No	No
Interlayer bond (with layer below)	Yes	Yes	No

[a] If the binder course is trafficked before the surface course is laid, otherwise 'No'.

[b] 'No' for surface deformation resistance but adequate stiffness needed to avoid structural deformation.

[c] For designed pavements, the stiffness of the surface course is often ignored for designing the pavement strength because future replacements may be different.

[d] Only in areas where studded tyres are used.

2.5 INTERNATIONAL SPECIFICATIONS

2.5.1 Harmonised European standards

The first editions of the European standard specifications for asphalt were published as the EN 13108 series in 2006 with a revision in 2016. The specifications were supported by a series of standard test methods in the EN 12697 series, the first of which had been published in 2000. Both series have been drafted in English, translated into French and German at each enquiry stage and, once published in those three languages, translated into other national languages. Prior to these standards, each country had its own standards for both specifications and test methods which differed markedly in terms of format, aspects covered and technical content, all of which were required to be withdrawn if their contents were superseded by the new standards. The requirement for harmonisation was intended to remove barriers to trade despite the fact that asphalt mixtures only cross national boundaries when both supplier and site are close to such a border. Nevertheless, the fact that the specifications are the same (if translated into a different language) means that suppliers and contractors should be able to understand job requirements more easily in all the countries across Europe that have adopted the standards. However, the use of different guidance documents for the use of the standards in specific countries does negate this advantage to a great extent. There are more European countries adopting CEN standards than just the European Union, with the complete list of such counties at the time of writing (before the implications of the June 2016 UK vote to leave the EU had been decided) being

- Austria
- Denmark
- Greece
- Latvia
- Norway
- Slovenia
- Czech Republic
- Belgium
- Estonia
- Hungary
- Lithuania
- Poland
- Spain
- FYR of Macedonia
- Bulgaria
- Finland
- Iceland
- Luxembourg
- Portugal
- Sweden
- Croatia
- France
- Ireland
- Malta
- Romania
- Switzerland
- United Kingdom
- Cyprus
- Germany
- Italy
- Netherlands
- Slovakia
- Turkey

There will be 11 parts of EN 13108 with an overall title of 'Bituminous mixtures – Material specifications'. The part titles are

- EN 13108-1, Asphalt concrete
- EN 13108-2, Asphalt concrete for very thin layers
- EN 13108-3, Soft asphalt
- EN 13108-4, Hot-rolled asphalt
- EN 13108-5, Stone mastic asphalt
- EN 13108-6, Mastic asphalt
- EN 13108-7, Porous asphalt
- EN 13108-8, Reclaimed asphalt
- EN 13108-9, Asphalt for ultrathin layers
- EN 13108-20, Type testing
- EN 13108-21, Factory production control

Parts 1–7 and 9 are specifications for different generic types of asphalt, Part 8 is a specification for a component material and Parts 20 and 21 are quality control standards.

Theoretically, no asphalt (or aggregate or bitumen) can be put onto the market in any of countries listed above without the product being CE Marked in accordance with the quality control standards plus one of the material specifications. Non-compliance has been made a trading standards issue as well as a contractual one with the potential for senior company officials being jailed for it.

The first editions of EN 13108 were drafted for hot mix asphalt, even though that assumption was not explicitly stated other than in the title for the associated test methods in the EN 12697, Bituminous mixtures – Test methods for hot mix asphalt series. However, lower temperature asphalts (Section 1.2) are being more widely used, which has created some confusion as to whether these materials are covered by EN 13108. In order to clarify matters, it has been decided to explicitly include them by

- Preparing a new part of EN 13108 for asphalt concrete with bitumen emulsion
- Clarifying the scope of each test method of EN 12697 as to which types of asphalt it is applicable during the next revision and then 'for hot mix asphalt' to be removed from the title

The parts of EN 12697 include sample preparations as well as actual test methods. The current parts are

- Part 1: Soluble binder content
- Part 2: Determination of particle size distribution
- Part 3: Binder recovery: rotary evaporator
- Part 4: Binder recovery: fractionating column
- Part 5: Determination of the maximum density
- Part 6: Determination of bulk density of bituminous specimens
- Part 7: Determination of the bulk density of bituminous specimens by gamma rays
- Part 8: Determination of voids characteristics of bituminous specimen
- Part 10: Compactibility
- Part 11: Determination of the affinity between aggregate and bitumen
- Part 12: Determination of the water sensitivity of bituminous specimens
- Part 13: Temperature measurement
- Part 14: Water content
- Part 15: Determination of the segregation sensitivity
- Part 16: Abrasion by studded tyres
- Part 17: Particle loss of porous asphalt specimen
- Part 18: Binder drainage
- Part 19: Permeability of specimen
- Part 20: Indentation using cube or cylindrical specimens
- Part 21: Indentation using plate specimens
- Part 22: Wheel tracking
- Part 23: Determination of the indirect tensile strength of bituminous specimens
- Part 24: Resistance to fatigue
- Part 25: Cyclic compression

- Part 26: Stiffness
- Part 27: Sampling
- Part 28: Preparation of samples for determining binder content, water content and grading
- Part 29: Determination of the dimensions of a bituminous specimen
- Part 30: Specimen preparation by impact compactor
- Part 31: Specimen preparation by gyratory compactor
- Part 32: Laboratory compaction of bituminous mixtures by vibratory compactor
- Part 33: Specimen prepared by roller compactor
- Part 34: Marshall test
- Part 35: Laboratory mixing
- Part 36: Determination of the thickness of a bituminous pavement
- Part 37: Hot sand test for the adhesivity of binder on pre-coated chippings for hot-rolled asphalt
- Part 38: Common equipment and calibration
- Part 39: Binder content by ignition
- Part 40: *In situ* drainability
- Part 41: Resistance to de-icing fluids
- Part 42: Analysis of coarse foreign matter in reclaimed asphalt
- Part 43: Resistance to fuel
- Part 44: Crack propagation by semi-circular bending test
- Part 45: Saturation ageing tensile stiffness (SATS) conditioning test
- Part 46: Low temperature cracking and properties by uniaxial tension tests
- Part 47: Determination of the ash content of natural asphalt
- Part 49: Determination of friction after polishing

It should be noted that Part 9 for reference density was withdrawn because the subject was covered by EN 13108-20. Other parts being developed at the time of writing are

- Part 48: Interlayer bond strength
- Part 50: Scuffing resistance of surface course
- Part 51: Surface shear strength test
- Part 52: Conditioning to address oxidative ageing
- Part 53: Cohesion increase by spreadability-meter method
- Part 54: Laboratory curing process for asphalt mixtures with bitumen emulsion

2.5.2 US Strategic Highway Research Program specifications

Toward the end of the last century, the United States was concerned about the condition of its road network. At the time, the Marshall and Hveem methods of mixture design were widely used but neither of these methods are performance based or even performance related. Therefore, a major national research was undertaken under the title of the Strategic Highway Research Program (SHRP) that produced the Superpave mixture design method and the associated performance grading of bitumen.

The objective of the Superpave mixture design system (Cominski, 1994) was to define an economical blend of bitumen and aggregate that produces an asphalt having: sufficient bitumen; sufficient voids in the mineral aggregate (VMA); a sufficient air voids contents; sufficient workability and satisfactory performance characteristics over the service life of the pavement.

The concept behind the Superpave mixture design system was to use available materials to prepare a mixture design that achieves a level of performance commensurate with the demands of traffic and environment on the pavement, the structure of the pavement and the reliability (i.e. minimisation of risk) required from the design. A flowchart of the Superpave mixture design produced is illustrated in Figure 2.1. The decisions at each step should be made so that the specific pavement performance requirements are met as far as practicable.

The three levels of design in the Superpave mixture design system permit the selection of a design process that is appropriate for the vehicle loads and traffic volume, measured in terms of total equivalent single axle loads (ESALs) of 80 kN over the service life of the pavement. Recommendations for applying the three design levels, which are suggested guidelines only, are reproduced in Table 2.3.

The design process increases significantly in complexity going from Level 1 to Level 3 with Level 3 requiring a greater number of tests, more test specimens and more time to complete a mixture design. However, the reliability of the design will increase with the extra effort so that the probability that the asphalt will perform satisfactorily for the pavement service life under the anticipated conditions of traffic and climate will also increase.

All three design levels explicitly consider the effects of the climate (environment) on pavement performance by selection of the performance grade (PG) of the bitumen. The performance grade is made up of the high and low pavement design temperatures in degrees Celsius at the site. As a general guide, if the two numbers differ by less than 90°C, the binder is unmodified bitumen while, if it is greater than that, it is a polymer-modified bitumen (PmB).

A design check list for use of Superpave (Brown et al., 2001) is

- Use a performance-graded (PG) binder and an N-design value appropriate for the weather, traffic level, and traffic speed for the project under consideration. Heavy, slow traffic will require a stiffer PG binder than may have been used in the past.
- Check that a complete mixture design has been undertaken in accordance with specifications and that it meets all of the aggregate consensus property requirements and specified volumetric criteria.

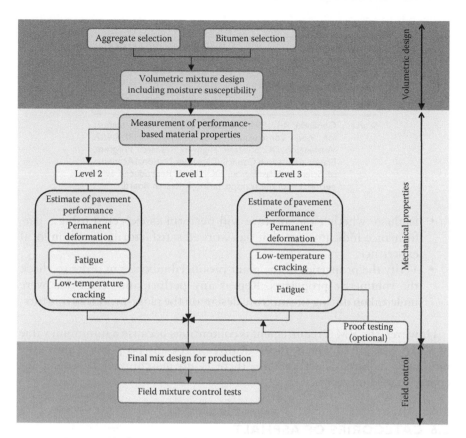

Figure 2.1 Structure of Superpave mixture design system. Level 3 provides the highest reliable estimate of pavement performance. (From Cominski, R J. 1994. The Superpave mix design manual for new construction and overlays. *SHRP-A-407*. Washington, DC: Strategic Highway Research Program, National Research Council; Copyright, National Academy of Sciences, Washington, DC, 1994. Reproduced with permission of the Transportation Research Board.)

- Check that the submitted design contains a reasonable bitumen content for the materials used and the design level specified.
- Generally, more filler (material passing the 0.075-mm sieve) is needed for coarse-graded mixtures. The character of the filler will control how much can be added to the mixture. Laboratory samples should contain the expected plant-produced amount of filler.
- In coarse-graded mixtures, if the VMA is more than 1.5% above the specified minimum, check for binder drainage.
- Excessive binder drainage is an indication that the bitumen content is too high for the bitumen grade, aggregate type and/or grading being used.

Table 2.3 Recommended design traffic levels

Design level	Design traffic (80 kN ESALs)
Level 1 (low)	$\leq 10^6$
Level 2 (intermediate)	$\leq 10^7$
Level 3 (high)	$> 10^7$

Source: Cominski, R J. 1994. The Superpave mix design manual for new construction and overlays. *SHRP-A-407*. Washington, DC: Strategic Highway Research Program, National Research Council; Copyright, National Academy of Sciences, Washington, DC, 1994. Reproduced with permission of the Transportation Research Board.

• Evaluate whether the mixture will perform as expected using a performance indicator test that has worked satisfactorily based on local experience.
• Verify the properties of the plant-produced mixture in order to check the volumetric properties. Repeat any performance tests that were undertaken during the mixture design on the plant-produced mixture.

However, the VMA requirement is contentious because a minimum value of VMA is required which, theoretically, can be any proportion above the minimum required. Furthermore, there is no matching requirement for coarse-graded mixtures.

2.6 CATEGORIES OF ASPHALT

2.6.1 Overview

There are various different types of asphalt that are generally thought of as

• Being defined by their aggregate grading and binder grade and quantity
• Having specific properties resulting from that composition

However, the specified gradings for each type of asphalt are not sufficiently unique to allow the intended mix type to be determined from an analysis of the mixture. This lack of clarity can be demonstrated by comparing the European target gradings for mixtures with a 10 mm maximum nominal-sized aggregate where a grading envelope can be classified as being for asphalt concrete (AC10), very thin layer asphalt concrete (BBTM10), stone mastic asphalt (SMA10) or porous asphalt (PA10), as shown in Figure 2.2.

Similarly, the different mixtures may have preponderance toward a 'good' or 'bad' value of specific properties, but the quantitative value

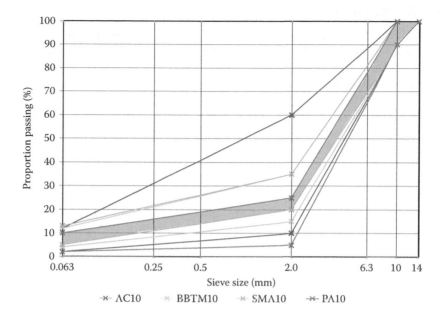

Figure 2.2 Overlapping aggregate grading envelopes.

of those properties can vary widely for each mixture type. A mixture to the combined grading would, it is assumed, be an open AC10, BBTM10 or SMA10 or a dense PA10 but with relative poor properties normally expected for each of these types.

The ideal approach for some engineers would be to replace specifying the asphalt mixture type by just specifying the properties required. However, many other engineers are more conservative because the ability to specify a mixture type allows the rest of the specification to be relatively limited for less sensitive jobs with low traffic. For heavily trafficked and/or more sensitive jobs, a fuller specification is still needed.

2.6.2 Asphalt concrete

2.6.2.1 General

Asphalt concrete (AC), also known as asphaltic concrete, is the basic type of hot mix asphalt that has been widely used around the world for many decades but has been developed into other mixture subtypes more recently. The European standard specification for AC is EN 13108-1 (CEN, 2016a), while the details for some of subtypes may be found in some national guidance documents, such as PD 6691 (BSI, 2015) for the United Kingdom. The products described below are such subtypes while more details about AC surface course mixtures can be found in Nicholls (1998).

2.6.2.2 Macadam

Macadam is a densely packed aggregate skeleton developed by Scottish engineer John Loudon McAdam around 1820 that, with a crowned profile, produced a relatively impermeable road prior to the development of bound pavement materials. The dense finish was produced by using a continuous grading such that each size of particles filled the gaps left in the skeleton of the larger sizes. Later, macadam was bound by tar, subsequently identified as a carcinogen, to produce tarmacadam or tarmac (the terms that the general public applies to all 'black' pavement materials, irrespective of grading or binder type) and later by bitumen to produce bitmac (later shortened to just macadam as the use of tar declined), these being bound layers.

Bound macadam has been the basic mixture for low-volume roads and as a structural layer for more major roads in the United Kingdom for a long time. The strength and porosity varies depending on the openness of the mixture, with the main classifications, as given in PD 6691 (BSI, 2015) being dense bitumen macadam or dense bituminous mixture (DBM), open-graded macadam, dense surface course and close-graded surface course. The binder for these mixtures is generally 40/60 bitumen although softer grades can also be used, particularly in colder areas. Variations of DBM have been developed to produce greater stiffness, these being heavy duty macadam (HDM) with an increased filler content and high modulus base (HMB), also called heavy duty mixture (HDM), with a stiffer binder, although the use of very stiff binders produced mixtures that were difficult to compact and subject to extremely premature deterioration and loss of the stiffness modulus for which it was developed or even complete disintegration (Figure 2.3).

2.6.2.3 Marshall asphalt

Bruce Marshall of the Mississippi Highway Department developed a mixture design method in 1939 which was refined by the US Army. Currently, the Marshall method is used in some capacity by about 38 American states. The principle of the method is to optimise the bitumen content in terms of the Marshall stability, Marshall flow, bulk density, voids in the total mixture and voids filled with bitumen. The first two properties were test methods developed specifically for the design method.

At one time, the Marshall design was the most popular method for designing asphalt when high performance was required. While asphalt design for highways has moved on with the Superpave method in the United States and the specific performance requirements for different properties, it is still widely used for airfield pavements. These AC mixtures designed to the Marshall method are called Marshall asphalt mixtures. Marshall asphalt can be used for all the bound layers of a pavement.

Marshall asphalt has a continuous grading that is held together by an optimised quantity of binder to produce a dense, impermeable mixture.

Figure 2.3 Poor integrity of HMB in the presence of water. (Courtesy of Dr. Mike E Nunn.)

Because of the higher quality control needed for Marshall asphalt compared to other mixtures, the UK Ministry of Defence (MoD, 2009) expects a dedicated mixing plant on site rather than a commercial plant which will produce a range of mixtures each working day.

2.6.2.4 Enrobé à module élevé

Enrobé à module élevé (EME) is an asphalt in which the binder is a particularly hard grade bitumen, generally 10/20 or 15/25, and has low air voids content. Therefore, EME is a very stiff mixture that has to be laid on stiff sub-base and cannot be exposed to oxidation, which would further embrittle the binder. Furthermore, it is only suitable for base and binder course because of the potential of binder ageing if used for the surface course.

EME was developed in France about 25 years ago as a binder course/base material that provides good mechanical properties (load spreading ability, resistance to deformation and cracking), durability and impermeability (Widyatmoko et al., 2007). The aggregate sizes for EME are 0/10, 0/14 and 0/20 mm. There are two classes of EME: EME Class 1 (EME1) with significantly lower binder content and less suitable for high-performance pavements and EME Class 2 (EME2) with a higher binder content and suitable for high-performance pavements.

2.6.2.5 Béton bitumineux pour chaussées aéronautiques

Béton bitumineux pour chaussées aéronautiques (AC for airfields, BBA) was developed in France, where it is the standard airfield asphalt surfacing mixture and has been used in almost all airport pavements in France. The

United Kingdom is now incorporating BBA into its standards. BBA can provide the required skid resistance without the need for grooving, allowing trafficking as soon as the material cools sufficiently.

There are four types of BBA materials that can be used for the binder course or the surface course in either a new construction or an overlay (Widyatmoko et al., 2007), these grades being 0/10 mm close graded, 0/10 mm gap graded, 0/20 mm close graded and 0/20 mm gap graded. Under the French specification, there are also three classes of BBA (i.e. BBA1, BBA2 and BBA3) with characteristics of mixture constituents, volumetrics and the level of performance tests required defined for different frequencies and weights of aircraft and for different climatic regions.

2.6.3 Porous asphalt

Porous asphalt (PA), previously known as pervious macadam, is a very open mixture with around 20% air voids content that is achieved by having predominantly a single size of aggregate particles. However, there does have to be some fine aggregate to form a mortar with the binder to hold the mixture together. The higher mortar content, particularly the binder content, will make the mixture more durable but can reduce its effectiveness for the particular properties that PA is used to achieve. The required binder contents are not capable of being held on the surface of the aggregate particles, with any extra binder draining off and dragging the other binder with it. The use of PmB increases the amount of binder that can be held because of its higher viscosity as can the inclusion of fibres of various forms because of the extra surface area.

The principal properties that make PA attractive are very low tyre/pavement noise generation, very low spray generation in the rain and avoidance of the buildup of water on the surface that can cause aquaplaning. However, these properties degenerate with time because of detritus filling the air voids. PA can be cleaned, but cleaning will not completely restore the original properties (Nicholls, 1997). However, the open nature of PA gives it poor structural properties and reduces the typical service life relative to more dense asphalts.

The European standard specification for PA is EN 13108-7 (CEN, 2016b).

2.6.4 Stone mastic asphalt

Stone mastic asphalt (SMA) was developed as a compromise between AC and mastic asphalt (MA) to overcome the problem of studded tyres on highways in Germany. The problem was later resolved by banning studded tyres, but the mixture had by then demonstrated its usefulness because of its robustness. It consists of a PA skeleton with the voids filled with a mortar of bitumen and filler. As with PA, fibres are used to carry the relatively high content of bitumen.

The PA skeleton provides good aggregate interlock that, in turn, gives good structural and deformation resistance. Therefore, it can be used in all of the bound layers of a pavement although it is not generally used as a base mixture in the United Kingdom.

The European standard specification for SMA is EN 13108-5 (CEN, 2016c).

2.6.5 AC for very thin layers

AC for very thin layers (béton bitumineux très minces, BBTM) is another type of asphalt that was developed in France. As the name suggests, BBTM is laid relatively thinly and, as such, is almost exclusively used for the surface course. BBTM is fairly similar to SMA except that PmB is used rather than fibres to carry the high binder content, which is actually higher than in SMA. However, distinguishing between the two mixture types by the use of fibres or PmB is complicated by the increasing use of both fibres and PmB for heavily stressed sites.

The European standard specification for BBTM is EN 13108-2 (CEN, 2016d).

2.6.6 Asphalt for ultrathin layer

Asphalt for ultrathin layer (AUTL) is yet another mixture type developed in France that is laid even thinner than BBTM and is used as the surface course. The mixture is almost single sized and is laid directly onto a bond coat distributed by a spray-bar that is fitted to the front of the paver. The water released from the emulsion as it breaks with the heat of the mixture can escape because of the thinness and open nature of the mixture when laid.

The European standard specification for AUTL is EN 13108-9 (CEN, 2016e).

2.6.7 Mastic asphalt

MA is a very dense mixture that does not contain coarse aggregates but only binder and the finer sizes of aggregate particles with a very high binder content of relatively hard bitumen. The strength of the mixture is from the bitumen rather than any aggregate interlock. The mixture is widely used for flat roofs as well as for pavements, generally bridge decks and footways. The use of hard bitumen requires high temperatures when laying. There are two types of MA:

- MA, which tends to be voidless and of the consistency of a pudding. It is hand-laid and generally used on bridge decks in the United Kingdom, France and the Mediterranean.

- Gussasphalt, which relies on a graded fine aggregate structure but, nevertheless, flows into place when assisted by compaction. It is used in Germany, Northern Europe and the Scandinavian countries.

Grip can be improved on MA by embedding chippings into the surface.

The European standard specification for MA is EN 13108-6 (CEN, 2016f).

2.6.8 Hot-rolled asphalt

Hot-rolled asphalt (HRA) was developed by Clifford Richardson in the United States under the name sheet asphalt (Richardson, 1905). HRA is formed from inserting coarse aggregate particles as 'plumbs' to bulk up a mastic of bitumen, filler and fine aggregate, although not with a particularly hard grade of bitumen. It is gap graded with very limited small coarse aggregate fractions in the grading. For highways, pre-coated chippings (PCCs) are often added to give better skid resistance. Without PCCs, texture will take time to develop. When PCCs are to be added, the proportion of coarse aggregate is generally 30% or 35% while, when PCCs are not to be added, the proportion is generally 80% or 85%.

The European standard specification for HRA is EN 13108-4 (CEN, 2016g).

2.6.9 Soft asphalt

Soft asphalt (SA) is a mixture type that is used in Scandinavia with relatively soft bitumen because of the lower ambient temperatures in that region. SA is primarily intended to be used for surface courses but can also be used for other bound layers of pavements in climates with low temperatures.

The European standard specification for SA is EN 13108-3 (CEN, 2016h).

2.6.10 Thin surface course systems

Thin surface course systems (TSCS), as used in the United Kingdom, are not another type of asphalt and not even necessarily thin. It would be more accurate to call them proprietary surface course systems because they are mixtures that have gone through a methodology to specify the properties that can be achieved by proprietary asphalt systems using a certification scheme. The definition of TSCS for the scheme (BBA, 2013) is that it

- 'has satisfactorily completed Stages 1 to 6 of this Guideline.
- can be installed at a nominal depth up to 50 mm. The actual depth range for the product will be defined by best practice based on the nominal size of the aggregate from the current version of BS 594987.

If depths outside this are requested, they will be confirmed via assessment.
• is a cold or hot bituminous based product'.

Therefore, any hot or cold material that includes bitumen and is not laid particularly thick (traditional surface course materials are generally laid 40, 45 or 50 mm thick) can become a TSCS if someone is willing to pay for the produce to go through the six stages of the scheme. The majority of TSCS are BBTM, AUTL or SMA mixtures.

REFERENCES

British Board of Agrément. 2013. *Interim Guideline Document for the Assessment and Certification of Thin Surfacing Systems for Highways*. Watford: British Board of Agrément. www.bbacerts.co.uk/wp-content/uploads/2014/10/Thin-Surface-Systems-Guideline.pdf

British Standards Institution. 2015. Guidance on the use of BS EN 13108, Bituminous mixtures – Material specifications. *PD 6691:2015*. London: BSI.

Brown, R, L Michael, E Dukatz, G Huber, R Sines and J Scherocman. 2001. Superpave mixture design guide. *WesTrack Forensic Team Consensus Report FHWA-RD-01-052*. Washington, DC: Federal Highway Administration.

Cominski, R J. 1994. The Superpave mix design manual for new construction and overlays. *SHRP-A-407*. Washington, DC: Strategic Highway Research Program, National Research Council.

Comité Européen de Normalisation. 2006. Bituminous mixtures – Material specifications – Part 1: Asphalt concrete. *EN 13108-1:2006*. London: BSI; Berlin: DIN; Paris: AFNOR; and other European standards institutions.

Comité Européen de Normalisation. 2016a. Bituminous mixtures – Material specifications – Part 1: Asphalt concrete. *EN 13108-1:2016*. London: BSI; Berlin: DIN; Paris: AFNOR; and other European standards institutions.

Comité Européen de Normalisation. 2016b. Bituminous mixtures – Material specifications – Part 7: Porous asphalt. *EN 13108-7:2016*. London: BSI; Berlin: DIN; Paris: AFNOR; and other European standards institutions.

Comité Européen de Normalisation. 2016c. Bituminous mixtures – Material specifications – Part 5: Stone mastic asphalt. *EN 13108-5:2016*. London: BSI; Berlin: DIN; Paris: AFNOR; and other European standards institutions.

Comité Européen de Normalisation. 2016d. Bituminous mixtures – Material specifications – Part 2: Asphalt concrete for very thin layers (BBTM). *EN 13108-2:2016*. London: BSI; Berlin: DIN; Paris: AFNOR; and other European standards institutions.

Comité Européen de Normalisation. 2016e. Bituminous mixtures – Material specifications – Part 9: Asphalt for ultra-thin layer (AUTL). *EN 13108-9:2016*. London: BSI; Berlin: DIN; Paris: AFNOR; and other European standards institutions.

Comité Européen de Normalisation. 2016f. Bituminous mixtures – Material specifications – Part 6: Mastic asphalt. *EN 13108-6:2016*. London: BSI; Berlin: DIN; Paris: AFNOR; and other European standards institutions.

Comité Européen de Normalisation. 2016g. Bituminous mixtures – Material speci-
fications – Part 4: Hot rolled asphalt. *EN 13108-4:2016*. London: BSI; Berlin:
DIN; Paris: AFNOR; and other European standards institutions.
Comité Européen de Normalisation. 2016h. Bituminous mixtures – Material
specifications – Part 3: Soft asphalt. *EN 13108-3:2016*. London: BSI; Berlin:
DIN; Paris: AFNOR; and other European standards institutions.
Gourdon, J L, J C Nicholls, A Pronk, T Kollanen, P Höbeda and R Leutner. 1999.
Effect of compaction methods on mechanical properties of bituminous mix-
tures (SPECOMPACT). *Final Report, Contract No. SMT4-PL97-1439.*
5th EC Framework programme.
Ministry of Defence. 2009. *Marshall Asphalt for Airfields*. London: HMSO.
https://www.gov.uk/government/uploads/system/uploads/attachment_data/
file/33545/spec_132009.pdf
Nicholls, J C. 1997. Review of UK porous asphalt trials. *TRL Report TRL264*.
Wokingham: TRL Limited.
Nicholls, J C (editor). 1998. *Asphalt Surfacings*. London: E & FN Spon.
Richardson, C. 1905. *The Modern Asphalt Pavement*. New York: John Wiley &
Sons. https://archive.org/details/modernasphaltpa00richgoog
Widyatmoko, I, B Hakim, C Fergusson and J Richardson. 2007. The use of
French asphalt materials in UK airfield pavements. In 23rd PIARC World
Road Congress Paris, 17–21 September 2007. Paris: PIARC – World Road
Association.   www.researchgate.net/publication/288824215_The_Use_of_
French_Airfield_Asphalt_Concrete_in_the_UK

Chapter 3

Composition

3.1 MIXTURE DESIGN AND COMPLIANCE REQUIREMENTS

Defining the composition of an asphalt mixture forms the principal requirements for a conventional specification or for recipe requirements in other specifications. However, they can also be part of a mixture design procedure, as in the Marshall design with binder content and voids in mineral aggregate, or as compliance requirements for a performance-based specification, particularly those based on laboratory tests. If the performance has been demonstrated for a particular asphalt mixture with specific components, it is generally simpler and quicker to check that the same mixture has been supplied than to repeat the performance tests.

Asphalt is not a particularly consistent building material because it is 'an inhomogeneous mixture of inhomogeneous natural component materials'. The actual performance can depend on many factors including the actual quarry (or even location within the quarry) from which the aggregate is taken, the method of crushing to produce the aggregate particles and the geographical source of the crude from which the bitumen is refined. Conventional specifications generally cannot be specific with regard to such issues, so checking the composition with fixed components is acceptable.

3.2 BINDER

3.2.1 Binder content

The simplest method to determine the binder content of an asphalt mixture is to measure the components that are added to make up that mixture. However, in most plants, the batching takes place in large quantities so that there may still be a need to check that the components have been distributed evenly to ensure that each load has the required binder content.

From asphalt mixtures, the binder needs to be extracted, the mineral aggregate needs to be separated and then the binder content calculated. However, there are several methods to carry out each of these operations,

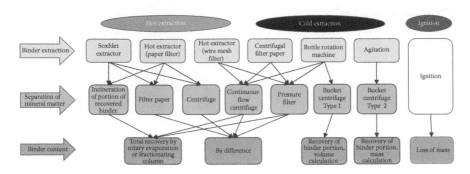

Figure 3.1 Alternative procedures for determination of binder content. (After Comité Européen de Normalisation. 2012a. Bituminous mixtures – Test methods for hot mix asphalt – Part 1: Soluble binder content. *EN 12697-1:2012*. London: BSI; Berlin: DIN; Paris: AFNOR; and other European standards institutions.)

as shown in Figure 3.1. Traditionally, these methods were solvent based as given in the European test method, EN 12697-1 (CEN, 2012a), in which the choice of solvent, which will have an effect on the result, is left open because it was found that each solvent was banned in at least one member state. The use of solvents is generally discouraged but the methods can be undertaken safely with appropriate precautions.

An alternative to the use of solvents that has been developed is the analysis by binder ignition, as standardised in EN 12697-39 (CEN, 2012b) for Europe. However, there is a potential difference between the 'soluble' binder and the 'flammable' solvent with materials including cellular fibres, which are insoluble and flammable, being classified as part of binder by the ignition method but not by solvent methods. Therefore, the test needs calibration for a mixture which will need to be repeated whenever there is a change in the composition, particularly in terms of the aggregate source. Other issues are that some aggregates can react at the high temperatures reached in the test and that the properties of the binder cannot be analysed after ignition, although the latter is also the case for some of the ignition methods.

The binder content is an important issue with regard to the performance of any asphalt mixture. However, there is not a universal rule that more, or less, is always better. Higher binder contents will improve the workability and durability of asphalts, but it will also reduce most mechanical properties (other than the resistance to fatigue) which will, practically, also reduce the durability.

The precision given in EN 12697-1 for binder content by solvent methods from three different experiments are repeatabilities of 0.23%, 0.23% and 0.3% and reproducibilities of 0.31%, 0.34% and 0.5% by mass for mixtures with paving grade bitumen while no data are given for mixtures with polymer-modified bitumen. The precision given in EN 12697-39 for binder content by ignition are a repeatability of 0.31% and a reproducibility of 0.56% with no reference to the binder. The data imply that the precision is marginally better for the solvent methods than the ignition method. Within the solvent methods, the methods using total recovery would be expected to be better than the methods by difference although no data are given to support that expectation.

For use in Europe, the tolerance required by EN 13108-21 (CEN, 2016i) on the binder content is ±0.5% for mastic asphalt mixtures and other mixtures apart from HRA with small aggregates and ±0.6% for HRA and other mixtures with large size aggregate. The precision values are sufficiently close to the tolerance to require multiple checks to be confident of (non-)compliance.

3.2.2 Binder properties

The main properties of the binder are its viscosity as measured under different conditions, generally penetration and softening point in Europe. However, another aspect is whether the bitumen has been modified and, if so, which modifier was used and in what concentration. There are a wide range of modifiers that are, or have been, used to modify bitumen as a binder with a number of examples shown in Table 3.1 (Nicholls, 1994).

Each modifier will modify the properties of the base bitumen differently.

The methods used to reclaim the binder from an asphalt mixture are usually either the rotary evaporator, as standardised in EN 12697-3 (CEN, 2013a) for Europe, and the fractionating column, as standardised in EN 12697-4 (CEN, 2015a) for Europe. The rotary evaporator method is suitable for the recovery of paving grade bitumen, while the fractionating column method is suitable for the recovery of volatile matter including cut-back bitumen. Polymer-modified binders can be treated in a similar manner, but often take longer to be fully extracted from the mixture with the extent of the extra time depending on the type of polymer and its concentration.

Test methods that are commonly used to measure the properties of the recovered binder are needle penetration, used as standardised in EN 1426 (CEN, 2015b) for Europe, and ring and ball softening point, as standardised in EN 1427 (CEN, 2015c) for Europe. However, these properties are often estimated using data from the dynamic shear rheometer (DSR) which requires a smaller sample than the 'traditional' test methods. The standard test methods for the DSR include EN 14770 (CEN, 2012c) in Europe and ASTM D7175 (ASTM, 2015a) in the United States.

The precision given in EN 1426 for penetration under standard conditions are

- A repeatability of 2 × 0.1 mm up to 50 pen and 4% above that value
- A reproducibility of 3 × 0.1 mm up to 50 pen and 6% above that value

Table 3.1 Types of bitumen modifier

Categories of modifiers	Examples of generic types	
Thermosetting polymer modifiers	Epoxy resin	Polyurethane resin
	Acrylic resin	
Thermoplastic polymer elastomers	Natural rubber	Vulcanised (tyre) rubber
	Styrene–butadiene–styrene block copolymer (SBS)	Ethylene–propylene–diene terpolymer (EPDM)
	Styrene–butadiene rubber (SBR)	Isobutene–isoprene copolymer (IIR)
Organic thermoplastic polymer modifiers	Ethylene–vinyl acetate (EVA)	Polypropylene (PP)
	Ethylene methyl acrylate (EMA)	Polyvinyl chloride (PVC)
	Ethylene butyl acrylate (EBA)	Polystyrene (PS)
	Polyethylene (PE)	
Chemical modifiers and extenders	Sulphur	Lignin
	Organo-manganese/cobalt compound	
Fibres	Cellulose	Asbestos
	Alumino-magnesium silicate	Polyester
	Glass fibre	Polypropylene
Anti-stripping	Organic amines	Amides
Antioxidants	Amines	Zinc antioxidant
	Phenolics	Lead antioxidant
Natural binders	(Trinidad) Lake asphalt	Uintaite (Gilsonite)
Fillers	Carbon black	Lime
	Fly ash	Hydrated lime

Source: Nicholls, J C. 1994. Generic types of binder modifier for bitumen. SCI Lecture Paper No 0035. London: Society of Chemical Industry.

The precision given in EN 1427 for softening point are

- A repeatability of 1.0°C and a reproducibility of 2.0°C for unmodified bitumen
- A repeatability of 1.5°C and a reproducibility of 3.5°C for polymer-modified bitumen

When using these properties to check the binder quality, changes in the binder properties from delivery to the batching plant need to be considered. The binder properties will have changed during the mixing and laying of asphalt, so that the recovered properties will not be identical to those of the fresh binder prior to those operations. Typically, the penetration will have reduced by about one grade and the softening point increased equivalently with the hardening of the binder involved. *In situ* ageing will further change the properties, but at a slower rate. For porous asphalt, which is the mixture which ages most because of the access to oxygen, the drop in

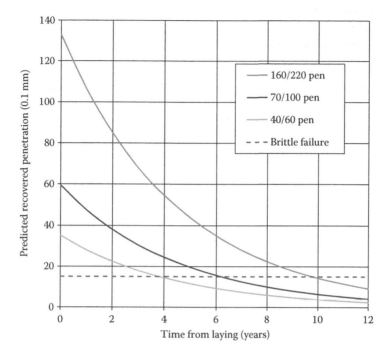

Figure 3.2 Predicted penetration of binder recovered from porous asphalt. (From Nicholls, J C. 1997. Review of UK porous asphalt trials. *TRL Report TRL264.* Wokingham: TRL Limited.)

penetration has been found (Nicholls, 1997) to be as shown in Figure 3.2. Brittle failure often occurs once the penetration has fallen below 15 pen.

A second potential influence on the binder properties is that any solvent used to extract the binder that has not been subsequently removed will soften the binder. The vast majority of the solvent should have been removed in a competent laboratory, but there may still be traces, particularly with polymer-modified binders that have unusual properties themselves.

3.3 AGGREGATE

3.3.1 Aggregate grading

The grading of the aggregate particles after the binder has been removed can be measured on sieves using methods such as EN 12697-2 (CEN, 2015d) in Europe. It is important that all the binder has been removed because any remaining binder will coarsen the apparent grading because any remaining binder will coarsen the apparent grading as a result of

- The binder film increasing with the particle radius by the binder film thickness
- The binder gluing the particles together, particularly the finer particles

The difference between the aggregate grading and the asphalt mixture grading is critical when large proportions of reclaimed asphalt are being used in a mixture. A calibration is needed to allow the aggregate grading to be accurately estimated from the reclaimed asphalt grading because bulk quantities of reclaimed asphalt cannot be economically stripped of binder for repeatedly measuring the grading of just the aggregate.

As with binder content, the aggregate grading is an important issue in the design of an asphalt mixture, but there is no ideal set of proportions for different gradings or else there would be no need for different types of asphalt. Nevertheless, the proportions do need to be controlled as far as practicable once properly designed to ensure that the design properties are not compromised.

The precision given in EN 12697-2 is a repeatability of 1.0% and reproducibility of 1.7% while the tolerance required in Europe by EN 13108-21 (CEN, 2016i) varies between ±2% for the filler and up to ±9% for the larger aggregate particles. When using these values to check compliance, it must be remembered that that mixtures from site are to be compared against the design grading and not the target composition grading envelope; if the design grading is along one edge of the target envelope, a proportion for a sieve size within the target envelope near the opposite edge could be outside the tolerance and, therefore, non-compliant while a proportion on the other side of the design grading by less than the tolerance would be outside the target envelope but compliant.

3.3.2 Particle shape

No two aggregate particles have identical shape, even when of the same nominal size. The most commonly used shape measurement is the flakiness index, which is standardised as EN 933-3 (CEN, 2012d) in Europe. There is also the shape index, which is standardised as EN 933-4 (CEN, 2008a) in Europe. When used, both tests are usually undertaken separately on each aggregate fraction. The precision in EN 933-3 for flakiness index are

- A repeatability of 1.9% and a reproducibility of 2.9% for a level of 10%
- A repeatability of 3.8% and a reproducibility of 8.2% for a level of 20%
- A repeatability of 2.2% and a reproducibility of 11.6% for a level of 50%

Generally, the flakiness index (and shape index) is kept low in order to achieve cubical or spherical particles that can fit together without excessive voids. However, flaky particles can be used successfully provided that they are not mixed with cubical or spherical particles of the same nominal and the flaky particles are oriented in the same plane. However, such mixtures will have different properties in different axes.

The limits on flakiness index in Europe set out in EN 13043 (CEN, 2002a) are for categories of less than 10% increasing by 5% to a category of less than 35% plus one for less than 50%, one for over 50% and one for no requirement. The limits for shape index in Europe are similar except there is no category for less than 10%. The categories to be set for different mixtures may be included in national guidance documents, with the UK guidance in PD 6691 (BSI, 2015) being a flakiness index category of less than 35% for all asphalt concrete mixtures except BBA mixtures, where it is less than 25%, and HRA and a flakiness index category of less than 20% for SMA and pre-coated chippings for HRA. PD 6691 gives no guidance on shape index.

3.4 VOLUMETRICS

3.4.1 Air voids content

Determining the air voids content involves the following simple calculation, as standardised in several test methods including EN 12697-8 (CEN, 2012e):

$$\text{Air voids content} = 100 \times \frac{(\text{maximum density} - \text{bulk density})}{\text{maximum density}} \% (\text{by volume})$$

Sometimes it has been referred to as a fundamental test but the air voids content found will depend on what is considered a void (Richardson and Nicholls, 1999). The reasons for possible differences include

- The aggregate particles in the asphalt can contain voids, some of which are connected to the exterior while others are closed, which may or may not be classed as voids in the asphalt.
- The surface texture and other voids at interfaces may or may not be regarded as voids in the asphalt.
- There will be differences between air voids, which are what is generally measured, and water voids, which is what is often the issue, because of the difference in surface tension and viscosity of the two fluids.

The definition of what is considered to be a void is covered by the procedures used for the measurement of maximum and bulk densities.

The European standard for maximum density, EN 12697-5 (CEN, 2009a), has three procedures:

- The volumetric procedure involves comparing the volume of the sample without voids and its dry mass. The volume of the sample is measured as the displacement of water or solvent by the sample in a pycnometer after it has been broken down to expose the air voids but without crushing aggregate particles.
- The hydrostatic procedure is similar but the volume of the sample is calculated from the dry mass and the mass in water of the sample, again after being broken down.
- The mathematical procedure involves calculating the contribution from the quantity and density of each of the constituent materials. The mathematical procedure was the traditional method, when it was known as maximum theoretical density.

The European standard for bulk density, EN 12697-6 (CEN, 2012f), has four procedures:

- The dry procedure involves comparing the weight in air to the weight in water and assumes that all accessible voids are excluded and, as such, is suitable for mixtures with few voids.
- The saturated surface dry (SSD) procedure is similar but the weighing in air is after being submerged following which any water on the surface is removed by blotting, so that only internal voids in which water remains are excluded and, therefore, is suitable for mixtures with a few more voids than the dry procedure.
- The sealed specimen procedure again compares the weight in air and the weight in water but the specimen is sealed prior to both weighings so that all voids enclosed are included in the results and is suitable for mixtures with many voids. The weight of the sealing matter has to be allowed for in the calculation.
- The procedure by dimensions compares the weight in air with the volume by measurement and is suitable for very open mixtures like porous asphalt.

Each method for maximum density is intended to produce the same numerical value although, in practice, they may differ slightly. Each method for bulk density measures a slightly different property and, therefore, is likely to produce a different numerical value for the same mixture that could be significant.

An alternative to these four laboratory procedures, each of which measures the overall bulk density of a specimen, is by means of a scan of a core with gamma rays to obtain the bulk density profile of that core, from which the air voids content can be determined. The European standard for this procedure is EN 12697-7 (CEN, 2014a), but it is generally used for investigations rather than specifications.

The density in itself is not particularly important because it depends heavily on the density of aggregate. However, the air voids content is

Table 3.2 Precision of density and air voids content measurements

Measure	Method	Standard	Units	Repeatability	Reproducibility
Bulk density	SSD	EN 12697-6	Mg/m³	0.017 + 0.0003 A	0.022 + 0.0006 A
	Gamma rays	EN 12697-7	Mg/m³	0.007	0.032
Maximum density	Using water	EN 12697-5	Mg/m³	0.011	0.015
	Using solvent		Mg/m³	0.019	0.042
Air voids content		EN 12697-8	% (v/v)	1.1	2.2

A = Proportion of aggregate particles passing the 11.2 mm sieve (%).

generally kept to a minimum (unless designing a pervious pavement) with a constraint of not making the mixture subject to permanent deformation at very low contents. The precision given in the European standards for density and air voids density are set out in Table 3.2.

The laboratory measurements for bulk density are disruptive as a measure of the *in situ* density because of the need to take cores. Nuclear density gauges were developed for such *in situ* checks but require calibration against a laboratory procedure for the particular mixture or mixtures. Care needs to be taken with the nuclear source, so alternative non-nuclear density gauges have also been developed using different techniques such as electromagnetic waves and nuclear gauges have been generally phased out. The non-nuclear gauges also need calibration against a laboratory procedure.

3.4.2 Voids filled with bitumen

The voids content in the mineral aggregate (VMA) of an asphalt mixture can be calculated from the air voids content, binder content and bulk density of the specimen plus the density of the binder. The voids filled with bitumen (VFB) (European terminology) or the voids filled with asphalt (VFA) (American terminology) in an asphalt mixture can be calculated from the binder content, voids in the mineral aggregate and bulk density of the specimen plus the density of the binder. Both formulae are given in standards, including EN 12697-8 (CEN, 2012e), with the VMA and VFB (or VFA) being used as design parameters in mixture design methods like the Marshall method.

3.4.3 Percentage refusal density

The percentage refusal density (PRD) is a comparison between the bulk density of a sample with the bulk density of that specimen after it has been reheated and re-compacted as much as possible (i.e. to refusal). The refusal density is not the same as the maximum density because the specimen has not been broken down and can still contain voids, as in the case of porous asphalt. The test is generally used to check that the *in situ* compaction has been effective.

3.5 TEMPERATURE

The temperature of the asphalt mixture is not really part of the composition, but it is dealt with here because it is a parameter that should be considered in the same way, particularly now that there are warm, semi-warm and cold as well as hot asphalt mixtures.

The standard method for measuring the temperature of the asphalt was traditionally the immersion of a thermometer within the uncompacted mixture. The difficulty with this type of measurement is the potential danger in inserting the thermometer into the mixture and the disruption caused by that action during the mixing, transporting and laying the asphalt. Nevertheless, the European standard, EN 12697-13 (CEN, 2003a), contains methods for measurements of the temperature of asphalt in a lorry, in laid material and in a heap but not in the paver. Several types of temperature measuring devices can be used, including thermocouples, thermistors and bimetallic rotary thermometers. EN 12697-13 gives no precision data.

The temperature can also be measured by remote equipment including infrared camera technology, which can be used to produce thermographic plots during paving (Figure 3.3). These images can be used to check for the consistency of the temperature and to check that the temperature is within a suitable range during compaction.

Figure 3.3 Example of thermographic plot of asphalt during paving. (Courtesy of Peter D Sanders.)

REFERENCES

ASTM International. 2015a. Standard test method for determining the rheological properties of asphalt binder using a dynamic shear rheometer. *ASTM D7175 – 15*. West Conshohocken, PA: ASTM International.

British Standards Institution. 2015. Guidance on the use of BS EN 13108, Bituminous mixtures – Material specifications. *PD 6691:2015*. London: BSI.

Comité Européen de Normalisation. 2002a. Aggregates for bituminous mixtures and surface treatments for roads, airfields and other trafficked areas. *EN 13043:2002*. London: BSI; Berlin: DIN; Paris: AFNOR; and other European standards institutions.

Comité Européen de Normalisation. 2003a. Bituminous mixtures – Test methods for hot mix asphalt – Part 13: Temperature measurement. *EN 12697-13:2003*. London: BSI; Berlin: DIN; Paris: AFNOR; and other European standards institutions.

Comité Européen de Normalisation. 2008a. Tests for geometrical properties of aggregates – Part 4: Determination of particle shape – Shape index. *EN 933-4:2008*. London: BSI; Berlin: DIN; Paris: AFNOR; and other European standards institutions.

Comité Européen de Normalisation. 2009a. Bituminous mixtures – Test methods – Part 5: Determination of the maximum density. *EN 12697-5:2009*. London: BSI; Berlin: DIN; Paris: AFNOR; and other European standards institutions.

Comité Européen de Normalisation. 2012a. Bituminous mixtures – Test methods for hot mix asphalt – Part 1: Soluble binder content. *EN 12697-1:2012*. London: BSI; Berlin: DIN; Paris: AFNOR; and other European standards institutions.

Comité Européen de Normalisation. 2012b. Bituminous mixtures – Test methods for hot mix asphalt – Part 39: Binder content by ignition. *EN 12697-39:2012*. London: BSI; Berlin: DIN; Paris: AFNOR; and other European standards institutions.

Comité Européen de Normalisation. 2012c. Bitumen and bituminous binders – Determination of complex shear modulus and phase angle – Dynamic shear rheometer (DSR). *EN 14770:2012*. London: BSI; Berlin: DIN; Paris: AFNOR; and other European standards institutions.

Comité Européen de Normalisation. 2012d. Tests for geometrical properties of aggregates – Part 3: Determination of particle shape – Flakiness index. *EN 933-3:2012*. London: BSI; Berlin: DIN; Paris: AFNOR; and other European standards institutions.

Comité Européen de Normalisation. 2012e. Bituminous mixtures – Test methods – Part 8: Determination of voids characteristics of bituminous specimen. *EN 12697-8:2012*. London: BSI; Berlin: DIN; Paris: AFNOR; and other European standards institutions.

Comité Européen de Normalisation. 2012f. Bituminous mixtures – Test methods – Part 6: Determination of bulk density of bituminous specimens. *EN 12697-6:2012*. London: BSI; Berlin: DIN; Paris: AFNOR; and other European standards institutions.

Comité Européen de Normalisation. 2003b. Road and airfield surface characteristics – Test methods – Part 7: Irregularity measurement of pavement courses – The straightedge test. *EN 13036-7:2003*. London: BSI; Berlin: DIN; Paris: AFNOR; and other European standards institutions.

Comité Européen de Normalisation. 2014a. Bituminous mixtures – Test methods for hot mix asphalt – Part 7: Determination of the bulk density of bituminous specimens by gamma rays. *EN 12697-49:2014*. London: BSI; Berlin: DIN; Paris: AFNOR; and other European standards institutions.

Comité Européen de Normalisation. 2015a. Bituminous mixtures – Test methods for hot mix asphalt – Part 4: Bitumen recovery – Fractionating column. *EN 12697-4:2015*. London: BSI; Berlin: DIN; Paris: AFNOR; and other European standards institutions.

Comité Européen de Normalisation. 2015b. Bitumen and bituminous binders – Determination of needle penetration. *EN 1426:2015*. London: BSI; Berlin: DIN; Paris: AFNOR; and other European standards institutions.

Comité Européen de Normalisation. 2015c. Bitumen and bituminous binders – Determination of softening point – Ring and Ball method. *EN 1427:2015*. London: BSI; Berlin: DIN; Paris: AFNOR; and other European standards institutions.

Comité Européen de Normalisation. 2015d. Bituminous mixtures – Test methods for hot mix asphalt – Part 2: Determination of particle size distribution. *EN 12697-2:2015*. London: BSI; Berlin: DIN; Paris: AFNOR; and other European standards institutions.

Comité Européen de Normalisation. 2016i. Bituminous mixtures – Material specifications – Part 21: Factory production control. *EN 13108-21:2016*. London: BSI; Berlin: DIN; Paris: AFNOR; and other European standards institutions.

Nicholls, J C. 1994. Generic types of binder modifier for bitumen. *SCI Lecture Paper No 0035*. London: Society of Chemical Industry.

Nicholls, J C. 1997. Review of UK porous asphalt trials. *TRL Report TRL264*. Wokingham: TRL Limited.

Richardson, J T G and J C Nicholls. 1999. Determination of air voids content of asphalt mixtures. In *3rd European Symposium on the Performance and Durability of Bituminous Materials*, University of Leeds. *TRL Paper PA 3418/98*. Zurich: Aedificatio.

Chapter 4

Surface characteristics

4.1 PROFILE

4.1.1 Level (fundamental)

The levels of the top surfaces of the various layers of the pavement are specified in order to ensure that the final structure interacts with the other structures (examples are buildings, overbridges, under-bridges and kerbs) appropriately. The accuracy needed is generally greater for higher layers because most of the interaction is at the surface, although the difference in levels between the top and bottom of any layer can be important in order to ensure that the thickness of the layer is sufficient to work as designed.

The level is a property resulting from the laying procedure rather than the material itself, although the nominal maximum size of the aggregate particles does affect the accuracy to which thinner layers can be laid – the accuracy cannot be too precise given that the surface of every level will not be perfectly planar but contain 'valleys' between the aggregate 'hills'.

The level is traditionally measured by surveying with dumpy levels, but can now be carried out with optical/laser levels and total stations, which are electronic theodolites (transits) integrated with electronic distance meters (EDM). Further details will not be given here because the level is not a material parameter.

4.1.2 Surface regularity (fundamental)

The quality of ride is dictated by what is described as the 'surface regularity' of the layer on which a vehicle is driven, usually the surface course. The quality of ride is better when the value of surface regularity is lower. For the pavement to retain a good ride quality, the pavement materials need to have sufficient deformation resistance (Section 5.2) to resist the traffic loading with due allowance for the expected improvement in deformation resistance as the material ages and becomes stiffer. The surface regularity is strongly allied to the levels at which the surface is laid (Section 4.1.1) but with a different emphasis and is also a property resulting from the laying process rather than the material itself. The levels can be compliant but any sharp changes in level within the tolerance can still lead to poor surface regularity.

The ideal for good ride quality would be a perfectly flat surface but the geography of most locations, apart from such areas as East Anglia in the United Kingdom, includes hills and valleys while the need for drainage usually requires cambers and/or crossfalls in the surface, particularly in flat areas. Nevertheless, the surface regularity can be improved if the paver is fitted with a sonic or other averaging beam.

The surface regularity is normally measured differently for longitudinal and transverse regularity. The traditional measurements are the rolling straightedge in the more continuous longitudinal direction and a 3-m-long beam and wedge in the transverse direction although other methods of measurements are available with more being developed. However, because the initial surface regularity is not a material parameter and the maintenance of the parameter is covered by the deformation resistance, further details will not be given here.

The European test method for the straightedge is EN 13036-7 (CEN, 2003b) which uses a 3-m-long beam with no stand, which means that, when placed, the wedge can rest with gaps under the ends. The scope clarifies that the method is not applicable for determining the overall profile or general unevenness but for measuring single irregularities which are, by their nature, random.

Transverse unevenness indices are defined in EN 13036-8 (CEN, 2008b) to encompass the cross-fall of the transverse profile, irregularities or different defects in the transverse profile (steps, ridges/dips and edge slumps) and the longitudinal ruts in the wheel paths caused by the traffic.

4.2 SKID RESISTANCE

4.2.1 Friction

A major requirement of the surface of a pavement is to provide friction in order to enable vehicle tyres to grip the road, especially when braking and/or cornering. Ideally, the wheels will continue to turn so that the vehicle's kinetic energy can be converted into heat in the braking system, which is particularly important in wet conditions. Although vehicle tyres play an important part, the surface itself makes a significant contribution to the road friction. It is important to realise that friction is not constant, but is dependent on the nature of the surfacing, weather conditions, time of year and vehicle speed. When the road is wet, its skidding resistance can fall by more than 50% from its dry condition and, as vehicle speed increases, it falls further still. The essential components that the surface course contributes to friction are the texture depth (also known as macro-texture) of the surface and the micro-texture of the coarse aggregate within the surface.

4.2.2 Texture depth

4.2.2.1 Relevance of texture depth

Texture depth, or macro-texture, results from the coarser spaces between particles or grooves in the surface of the pavement (Figure 4.1). Texture depth contributes to the water drainage, which is important to allow any water to be able to drain away from the tyre/road interface and enable the tyre to grip the micro-texture, which is particularly important if high-speed skidding resistance is to be maintained. The importance of removing water on the proportion of accidents is shown in Figure 4.2 (Roe et al., 1991), which illustrates the proportion of accidents on three road networks in the United Kingdom reported as occurring in wet and dry conditions and whether or not skidding was involved, with about 50% of accidents in the wet involving skidding but only about 30% in the dry.

Research has also shown both that the accident risk increases with reduced texture depth, even in dry conditions (Roe et al., 1991) and that the texture depth has a strong influence on the relationship between skidding resistance and speed, particularly at low speeds (Roe et al., 1998).

There are several methods to measure texture depth, with the two principle methods defining texture depths differently.

4.2.2.2 Patch test (surrogate)

The patch test involves spreading a fixed volume of fine material onto the pavement and then using a spreader to create a roughly circular patch with a smooth surface where any voids in the pavement surface are filled by the particles. From the diameter of the circle and the volume of material, the average depth of the voids can be calculated as the texture depth. The volume of material can be reduced for smoother surfaces in order to limit the size of the patch.

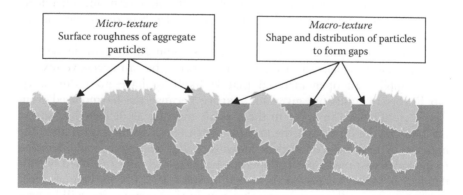

Figure 4.1 Schematic of micro- and macro-texture.

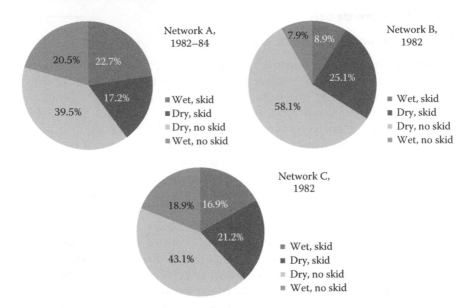

Figure 4.2 Influence of wet conditions on accidents on three road networks. (From Roe, P G, D C Webster and G West. 1991. The relation between the surface texture of roads and accidents. *TRL Research Report RR296*. Wokingham: TRL Limited.)

The fine material was traditionally a fixed grading of sand (when the test was known as the sand-patch test), but in the European test EN 13036-1 (CEN, 2010a) and the American test ASTM E965 (ASTM, 2015b), the sand is replaced by glass beads. The results are not affected significantly by the change (Nicholls et al., 2006a), but there has been a reluctance to abandon sand for the more expensive and slippery (at least locally for a short time) glass beads on the pavement surface.

The patch test is a manual test that has to be undertaken with an operative present on the pavement, requiring traffic management. Therefore, the test tends to be used for the initial texture depth prior to opening the pavement to traffic.

The texture depth is a property that is required to provide friction between the pavement surface and vehicle tyres, but an excessive texture depth implies an open material that is likely to have poor durability. No European limits are given for texture because the harmonised standards are intended to apply to the asphalt mixture as delivered to site in the back of the lorry, and the texture depth is not formed until after the material is laid and compacted. However, that argument could be used to exclude most mechanical properties because the values reached will be dependent on the compaction provided; nevertheless, EN 13108 gives various categories on the assumption that appropriate compaction will be provided. The UK national requirements for texture (HA et al., 2008a), which are relatively high because of the UK concern for safety, is a minimum of 0.9 mm or 1.5 mm depending on the situation but with plans to introduce a maximum requirement as well. The precision given in EN 13036-1 for the patch texture depth is a repeatability of 0.166 mm and a reproducibility of 0.321 mm.

4.2.2.3 Sensor-measured texture depth (surrogate)

Sensor-measured texture depth (SMTD) devices define the texture depth as the root mean square of the distance from an arbitrary plane to the pavement surface texture from a sequence of displacement measurements using a laser. The texture depth by SMTD is the standard deviation of the depth compared to average of the patch method. The two values have been correlated (Austroads, 2011), but the reliability of any correlation is dependent on the type of surface and, hence, the coefficient of variation.

The devices are mounted in vehicles, either as an attachment to a sideway-force coefficient routine investigation machine (SCRIM) or as part of dedicated multi-property measuring vehicle including the HARRIS and SCANNER vehicles in the United Kingdom. The advantage of the measurement is that it can be measured at traffic speed without requiring temporary closures.

4.2.2.4 Mean profile depth (surrogate)

The mean profile depth (MPD) is another measure of the macro-texture that defines it as the height of the highest peak above the mean level, which is different to the definition for both the patch method (mean texture depth) and the SMTD method (root mean square of the depth). Therefore, it is not surprising that these measures do not produce identical results, with the relationship between SMTP and MPD measurements from a large sample of rural roads being shown in Figure 4.3 (Viner et al., 2006), which is non-linear. MPD is determined as the mean of two peak levels minus the mean profile level as measured by a laser, other electrooptical device or sound transmission sensor. It can be used on laboratory sample or *in situ* where, unlike the patch method, it does not require a road closure. The MPD has been standardised internationally as ISO 13473-1 (ISO, 2004) and by the Americans as ASTM E1845-15 (ASTM, 2015c).

No precision data are given for the MPD test in ISO 13473-1.

4.2.2.5 Pavement surface horizontal drainability

The pavement surface horizontal drainability is a method for estimating the texture of smooth, non-porous surfaces (<0.4 mm MPD) by determining the horizontal drainability using an outflow meter as a stationary device. The method is standardised in Europe as EN 13036-3 (CEN, 2002b).

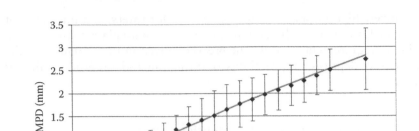

Figure 4.3 Relationship between SMPT and MPD on sampled rural roads. (From Viner, H et al. 2006. Surface texture measurement on local roads. *TRL Published Project Report PPR148*. Wokingham: TRL.)

4.2.3 Micro-texture

4.2.3.1 Relevance of micro-texture

The micro-texture is the fine texture of aggregate particles at the pavement surface (Figure 4.1) that break through the water film in wet conditions and, hence, allows frictional forces to be generated. Micro-texture is the key component in road surface friction without which the road would have hardly any friction in the wet. However, micro-texture can be polished away by traffic, particularly heavy traffic, so that the aggregate type needs to have both the micro-texture and the ability to resist wear. Research (Roe and Hartshorne, 1998) has led to changes in the requirements for the polishing resistance of aggregates in new surfaces.

Micro-texture has traditionally been measured by the polished stone value (PSV) of the aggregate with the resistance to wear by the aggregate abrasion value (AAV) but the friction after polishing (FAP) test has recently been introduced as a possible alternative.

4.2.3.2 PSV and AAV (surrogates on component)

PSV is a measure of the resistance of coarse aggregate to the polishing action of vehicle tyres under conditions similar to those occurring on the surface of a road. Single-sized 10/7.2 mm aggregate particles are glued onto a curved plate before being polished in an accelerated polishing machine with sand and water being fed across the face of the plate. The skid resistance of the aggregate is then determined using a pendulum friction tester.

The associated AAV is not a measure of skid resistance but of whether the skid resistance can be quickly abraded away. The test is undertaken on 14/10.2 mm aggregate particles embedded in resin on a curved plate that is fixed in contact with a horizontally rotating lap wheel. The wheel is rotated with an abrasive fine aggregate (sand) fed continuously through the contacting surfaces of the specimen and lap wheel for a specified number of revolutions. The AAV is determined from the differences in mass of the specimens before and after abrasion.

Both these tests are effectively conventional requirements because they are tests on a component material rather than on the mixture. In Europe, the standard test for PSV is given in EN 1097-8 (CEN, 2009b) with that for AAV being given in an annex of the same standard. EN 13043 (CEN, 2002a) gives the categories of PSV that can be specified in Europe which are not less than 44, 50, 56, 62 and 68 PSV units plus a declared and a no requirement categories in while the AAV categories are up to 10, 15 and 20 AAV units plus a declared and a no requirement categories.

A high PSV value implies improved pavement/tyre skid resistance. The minimum values necessary depend on the situation of the road, as shown by the requirements demand of UK trunk roads reproduced in Table 4.1 from HD 36/06 (HA et al., 2006a). The associated maximum values of AAV vary between 10 and 16 depending on the traffic flow and material type.

The precision of the PSV test is given as a repeatability of 0.042 times the value and a reproducibility of 3.30 plus 0.0333 times the value while the precision of the AAV are given as repeatabilities of 0.6 and 2.50 and reproducibilities of 1.5 and 3.5 for AAV values of 4 and 16, respectively.

4.2.3.3 Friction after polishing (simulative)

FAP can be measured on samples either of aggregate particles or of asphalt mixtures, including *in situ* samples taken by coring. The samples are polished with three polishing rollers that can be lowered and that move across the test surface at a predefined loading force whilst being continuously supplied with a mixture of water and quartz powder. After polishing, a separate rotating measuring head with three rubber sliding blocks is lowered onto the test surface while water is being added. The moment generated by the contact between the rubber sliders, after being de-clutched electronically, and the surface is continuously measured and recorded until the measuring head comes to a standstill with the friction being defined as the moment measured at 60 km/h.

Despite the available equipment required for the test not all being identical despite all coming from the same manufacturer, Wehner–Schulze, the FAP has been standardised for asphalt surfaces as EN 12697-49 (CEN, 2014b). Meanwhile, research is continuing across Europe on the capabilities of the FAP equipment (Daskova and Kudrna, 2012; Dunford and Roe, 2012a,b; Friel et al., 2013).

Although the test had only been recently standardised, the 2016 versions of EN 13108 includes categories for the results from asphalt concrete,

Table 4.1 Minimum PSV of surface course aggregates for UK trunk roads

Site cate-gory	Site description	Investigatory level	Minimum PSV required Traffic (cv/lane/day) at design life									
			0–250	251–500	501–750	751–1,000	1,001–2,000	2,001–3,000	3,001–4,000	4,001–5,000	5,001–6,000	Over 6,000
A1	Motorways where some braking regularly occurs (e.g. on 300 m approach to an off-slip)	0.30	50	50	50	50	50	55	55	60	65	65
		0.35	50	50	50	50	50	60	60	60	65	65
A2	Motorways where some braking regularly occurs (e.g. on 300 m approach to an off-slip)	0.35	50	50	50	55	55	60	60	65	65	65
B1	Dual carriageways where some braking regularly occurs (e.g. on 300 m approach to an off-slip)	0.30	50	50	50	50	50	55	55	60	65	65
		0.35	50	50	50	50	50	60	60	60	65	65
		0.40	50	50	50	55	60	65	65	65	65	68+
B2	Dual carriageways where some braking regularly occurs (e.g. on 300 m approach to an off-slip)	0.35	50	50	50	55	60	65	65	65	65	68+
		0.40	55	60	60	65	65	68+	68+	68+	68+	68+
C	Single carriageways where traffic is generally free-flowing on a relatively straight line	0.45	50	50	50	55	55	60	60	65	65	65
		0.50	55	60	60	65	65	68+	68+	68+	68+	68+
		0.55	60	60	65	65	68+	68+	68+	68+	68+	68+

(Continued)

Table 4.1 (Continued) Minimum PSV of surface course aggregates for UK trunk roads

| | | | Minimum PSV required | | | | | | | | | |
| | | | Traffic (cv/lane/day) at design life | | | | | | | | | |
Site category	Site description	Investigatory level	0–250	251–500	501–750	751–1,000	1,001–2,000	2,001–3,000	3,001–4,000	4,001–5,000	5,001–6,000	Over 6,000
G1/G2	Gradients >5% longer than 50 m	0.45	55	60	60	65	65	68+	68+	68+	68+	HFS
		0.50	60	68+	68+	HFS	HFS	HFS	HFS	HFS	HFS	HFS
		0.55	68+	HFS	HFS	HFS	HFS	HFS	HFS	HFS	HFS	HFS
K	Approaches to pedestrian crossings and other high-risk situations	0.50	65	65	65	68+	68+	68+	HFS	HFS	HFS	HFS
		0.55	68+	68+	HFS	HFS	HFS	HFS	HFS	HFS	HFS	HFS
Q	Approaches to major and minor junctions on dual carriageways and single carriageways where frequent or sudden braking occurs but in a generally straight line	0.45	60	65	65	68+	68+	68+	68+	68+	68+	HFS
		0.50	65	65	65	68+	68+	68+	HFS	HFS	HFS	HFS
		0.55	68+	68+	HFS	HFS	HFS	HFS	HFS	HFS	HFS	HFS
R	Roundabout circulation area	0.45	50	55	60	60	65	65	68+	68+	HFS	HFS
		0.50	68+	68+	68+	HFS	HFS	HFS	HFS	HFS	HFS	HFS

(Continued)

Table 4.1 (Continued) Minimum PSV of surface course aggregates for UK trunk roads

Site category	Site description	Investigatory level	Minimum PSV required									
			Traffic (cv/lane/day) at design life									
			0–250	251–500	501–750	751–1,000	1,001–2,000	2,001–3,000	3,001–4,000	4,001–5,000	5,001–6,000	Over 6,000
S1/S2	Bends (radius <500 m) on all types of road, including motorway link roads; other hazards that require combined braking and cornering	0.45	50	55	60	60	65	65	68+	68+	HFS	HFS
		0.50	68+	68+	68+	HFS	HFS	HFS	HFS	HFS	HFS	HFS
		0.55	HFS	HFS	HFS	HFS	HFS	HFS	HFS	HFS	HFS	HFS

Source: The Highways Agency, Transport Scotland, Welsh Assembly Government and The Department for Regional Development, Northern Ireland. 2006a. Surfacing materials for new and maintenance construction. In Design Manual for Roads and Bridges: Volume 7, Pavement Design and Maintenance: Section 5, Surfacing and Surfacing Materials: Part 1, HD 36/06. London: The Stationery Office. www.standardsforhighways.co.uk/dmrb/vol7/section5/hd3606.pdf

Note: Site categories are grouped according to their general character and traffic behaviour. The investigatory levels (IL) for specific categories of site are defined elsewhere. The IL to be used here must be that which has been allocated to the specific site on which the material is to be laid; Motorway or dual carriageway slip roads may fit in a number of groups depending on their layout; Where '68+' material is listed in this table, none of the three most recent results from consecutive PSV tests relating to the aggregate to be supplied must fall below 68; HFS means specialised high friction surfacing, incorporating calcined bauxite aggregate, will be required; For site categories G1/G2, S1/S2 and R any PSV in the range given for each traffic level may be used for any IL and should be chosen on the basis of local experience of material performance. In the absence of this information, the values given for the appropriate IL and traffic level must be used; Where designers are knowledgeable or have other experience of particular site conditions, an alternative PSV value can be specified.

BBTM, SMA, porous asphalt and AULT with categories having minima FAP values of 0.30 rising by 0.02 increments to 0.50 plus a no requirement category. The precision given for the FAP test, measured on BBTM specimen, are

- A repeatability of 0.026 and a reproducibility of 0.052 before polishing
- A repeatability of 0.024 and a reproducibility of 0.074 after 180,000 passes

However, there is some concern that the only manufacturer of the equipment still makes minor modifications to the design.

4.2.4 *In situ* measurement

4.2.4.1 *Variation of* in situ *skid resistance*

The skid resistance of a road is generally at its highest on installation. Traffic will polish the road surface from then, although its skid resistance does reach an 'equilibrium' level in about 2 years, the level depending on the type and speed of the traffic. Even after reaching equilibrium, the skid resistance will vary throughout the year with differing weathering conditions. Under UK weather conditions, it is lowest in summer and usually increases to some extent during the winter, presumably with the additional rainfall washing away some of the grit that erodes the aggregate. Given these variations:

- Results for the first couple of years may be greater than the equilibrium value
- Measurements have to be corrected for the time of year made, either by averaging several values or by a correction factor

Skid resistance is reduced in wet conditions, so that there are controlled amounts of water for each test. Furthermore, the skid resistance varies with vehicle speed and, therefore, this factor also has to be standardised for network monitoring.

4.2.4.2 *Pendulum measurements (simulative)*

The simplest method for measuring skid resistance is with a pendulum. The pendulum grazes over the wetted surface of the pavement and the friction is determined as to how far up the pendulum rises on the other side. The test is only suitable for surfaces not containing large-sized aggregate particles and the measurement is a manual one and is, therefore, not appropriate for network monitoring. However, it can be used for laboratory, as well as *in situ*, measurements. The test is standardised in Europe as EN 13036-4

(CEN, 2011) with a precision given as a repeatability of 6.648 and reproducibility of 7.202.

4.2.4.3 Sideway-force measurements (simulative)

A rotating wheel on the test vehicle is set at an angle to the direction of travel and a controlled amount of water spread on the road surface just in front of the wheel. The force developed on the angled wheel is measured to give a 'sideway-force coefficient' (SFC). The test tyre rotates slower than the vehicle tyres so that the measurements are for that slower speed. However, the rate at which the rubber in the contact patch slides in the forward direction is always significantly less than the speed of the test vehicle so the wheels wear comparatively slowly. An example of equipment using this technique is the SCRIM as shown in Figure 4.4, which is routinely used on the main road networks in the United Kingdom.

4.2.4.4 Fixed-slip measurements (simulative)

A tyre, with a controlled amount of water spread on the road surface just in front of it, running in the direction of travel, is forced to rotate more slowly than required for the speed at which the equipment is travelling and, hence, is constantly slipping. The force required to slow that tyre is the fixed-slip measure of skid resistance. Again, the actual slip speed of the tyre is slower than the speed of the equipment. An example of equipment for this method is the GripTester, which can be either moved manually as shown in Figure 4.5 (which further reduces the measurement speed) or towed behind a vehicle.

Figure 4.4 Sideway-force coefficient routine investigation machine. (Courtesy of John Prime.)

Figure 4.5 GripTester. (Courtesy of John Prime.)

4.2.4.5 Locked-wheel measurements (simulative)

The equipment travels at the measure speed when a wheel is locked so as to be sliding but not rotating while a controlled amount of water spread on the road surface just in front of it. The force generated is measured to determine the locked-wheel skid resistance for which the slip speed is the same as the speed of the test vehicle. The locked-wheel approach is the only one which can measure sliding friction directly over a wide range of speeds but the method is not suitable for measurements at higher speeds without traffic controls. An example of locked-wheel equipment is the Pavement Friction Tester (PFT) as shown in Figure 4.6.

4.2.5 Specifying skid resistance

The ideal would be to specify the equilibrium skid resistance, but that requires waiting two years before it can be assessed properly. Even measurements of initial skid resistance cannot be made until the works have been completed and any assessment of the equilibrium skid resistance from that is uncertain. Any corrective action required because of insufficient skid resistance after that time would be expensive and disruptive. Therefore, the surrogates of minimum texture depth and PSV with the associated maximum AAV are generally specified for new construction, although the PSV may be replaced by FAP once there is sufficient experience with the test. However, there are now moves to change to an acceptable range for the texture depth in order to ensure that the texture is not too open for good durability.

Figure 4.6 Pavement friction tester. (Courtesy of John Prime.)

The pavement surface horizontal drainability is only appropriate for assessing the texture depth of smooth surfaces, so is not appropriate for countries like the United Kingdom that require high levels of road safety. The measurement of texture depth is usually by the patch method at the time of construction, but the various methods that can be done at traffic speed such as SMPT and MPD are more appropriate for checking the maintenance of that texture.

Once the pavement has reached its equilibrium level, it would seem logical to monitor the *in situ* skid resistance. However, they each measure the skid resistance at different speeds and other constraints so none of them give the overall picture. Generally, the method required will be based on the equipment available to measure it.

4.3 NOISE AND SPRAY REDUCTION

4.3.1 Noise reduction

4.3.1.1 Issue being addressed

Road traffic noise is perceived as a problem by the public, whether travelling on, living near to or otherwise in the vicinity of roads. This perception has increased as vehicles have become quieter and expectations have increased. The properties of the pavement surface of texture depth, flow resistivity and acoustic absorption affect the generation and propagation of tyre/road noise. The other aspect of road noise is the engine noise with the relative proportion of the two sources varying depending on the composition of the traffic and weather conditions as well as the properties of the pavement.

4.3.1.2 Statistical pass-by method (fundamental)

The most commonly used method of measuring the relative noise perceived on reduction for different roads is the statistical pass-by (SPB) method, which has been standardised in ISO 11819-1:2001 (ISO, 2001). In the SPB method, the maximum A-weighted sound pressure levels and vehicle speeds of a statistically significant number of individual vehicles are measured as they pass a specific location. The vehicles of interest are cars, dual-axle heavy vehicles and multi-axle heavy vehicles. The sound pressures and logarithm of speed are plotted for each vehicle category and speed category (45–64 km/h for low; 65–99 km/h for medium and 100 km/h or more for high) and trend lines fitted so that the vehicle sound level can be determined for the vehicle and speed categories as the average maximum A-weighted sound pressure. The vehicle sound levels for the different vehicle types are added on a power basis based on the expected proportions in the flow to produce the statistical pass-by index (SPBI). No precision data are given for the test.

This test method is fundamental because it measures the noise that is generated on site. The main problem with it is that it can only be used to measure the noise when the site is suitable in order to avoid contamination from other sources of noise and/or reflection of the noise.

In the United Kingdom, the levels of noise reduction relative to 'traditional' surface course materials using the SPB are given Table NG 9/30 of the Notes for Guidance on the Specification for Highway Works (HA et al., 2008b), reproduced as Table 4.2.

A variant of SPB, where the microphone is mounted on a backing board rather than being used in normal free-field conditions, has also been standardised as ISO/PAS 11819-4 (ISO, 2013). By mounting the

Table 4.2 Road/tyre noise levels

Level	Equivalence to traditional surface course materials	Road surface influence (RSI)
3	Very quiet surface course material	−3.5 dB(A)
2	Quieter than HRA surface course materials	−2.5 dB(A)
I	Equivalent to HRA surface course materials	−0.5 dB(A)
0	No requirement	No requirement

Source: The Highways Agency, Transport Scotland, Welsh Assembly Government and The Department for Regional Development, Northern Ireland. 2008b. Notes for guidance on the specification for highway works, series NG 900, road pavements – Bituminous bound materials. In *Manual of Contract Documents for Highway Works, Volume 2*. London: The Stationery Office. www.standardsforhighways.co.uk/ha/standards/mchw/vol2/pdfs/series_ng_0900.pdf

microphone membrane very close to the backing board during measurements, noise from behind including reflections from facades or noise barriers is suppressed. The noise coming from the front is reflected by the backing board in a controlled way so that it can be taken into account by applying a correction to the measured value. Theoretically, the results increase by 6 dB because of the doubling of the sound pressure with the backing board.

4.3.1.3 Close proximity method (fundamental)

The problem with the SPB method is that the measurements have to be taken at a fixed distance from the vehicles with no impediments around that could muffle or reflect the noise. This requirement significantly limits the sites where measurements can be made.

In order to overcome the disadvantage of needing a suitable site, measurement can be made on a vehicle that is either self-powered or towed behind another vehicle. However, the draft for such a close proximity (CPX) method for measuring road noise, ISO DIS 11819-2, has not been finalised at the time of writing.

The CPX method involves measuring the average A-weighted sound pressure levels emitted by a specific tyre at a known traffic speed over an arbitrary distance of travel by at least two microphones located close to the tyre. Two uniquely different reference tyres are to be defined in order to represent the tyre/road characteristics. A substantial part of the propagation effect by acoustically absorptive surfaces should be represented in the microphone signal because the source of tyre/road noise is in CPX to the tyre/road interface. Measurements are made at the nominated reference speeds of 40 km/h, 50 km/h, 80 km/h and/or 100 km/h. Again, this test method is fundamental because it measures the noise that is generated on site.

4.3.1.4 Specifying for noise

Because there is no standardised method that can be used on any site, the only way to specify for noise is to use suitable sites as reference for a particular mixture using the SPB method with or without a backing board. The problem with this approach is the variability that will occur in the production and, probably more importantly, the laying of any asphalt mixture can affect the actual performance of the finished pavements. Furthermore, the pavement/tyre noise will tend to increase with use, particularly if there is any aggregate loss. Whilst it does have the advantage that a new pavement should always be quieter than the one it replaces (provided the mixture type is not change), it does make it difficult to maintain the noise performance with time.

4.3.2 Splash and spray reduction (surrogate)

Splash and spray can be a problem on highways during rainfall because of the reduced visibility from droplets rebounding from the surface or being thrown up by the rotating tyres; there is less of a problem for airports because pilots generally sit higher than car and lorry drivers.

Porous asphalt is largely considered to be the best material for reducing the splash and spray because the water percolates through the interconnecting voids. However, research on drainage systems (Nicholls and Carswell, 2001) found that there was a film of water at the surface of new porous asphalt even with relative modest rain. Nevertheless, the effectiveness is due to the water being pushed down into the material as tyres pass, even if it just reemerges nearby.

In situ drainability (also known as hydraulic conductivity) to EN 12697-40 (CEN, 2012g) has been used as a surrogate for splash and spray reduction, particularly for porous asphalt surface courses. The test uses a permeameter to determine the outflow time for a fixed quantity of water to dissipate through an annular area into the surface course of a pavement under known head conditions. The reciprocal of the outflow time is then used to calculate the relative hydraulic conductivity of the surface course material. The equivalent American test is ASTM D5084 (ASTM, 2010). These tests are not true permeability tests because they measure how quickly the water flows away rather than just through the pavement layer – much of the water will often reappear close to the permeameter. One problem for porous asphalt is that the surface course material will clog with detritus over time, with the hydraulic conductivity reducing significantly from when first laid (Nicholls, 1997).

Research has shown that the amount of splash and spray is influenced by the asphalt type and its properties, but it is also dependant on both the current and antecedent rainfall (Nicholls and Daines, 1992). There are some materials that will reduce the spray early in a rainstorm because there are 'holes' into which the rain can flow meaning that it does not have the potential to be propelled up to drivers' sight lines. However, later in the storm, those 'holes' will be full of water and the amount of spray will, if anything, be increased rather than reduced.

The main parameters, other than those of the pavement, that affect the splash and spray are generally considered (Sanders et al., 2012) to be water film thickness (which will depend on the geometry of the pavement surface plus both the current and antecedent rainfall), vehicle speed, tyre geometry, tyre tread depth, vehicle aerodynamics and any vehicle spray suppression devices such as flaps. This range of properties makes it difficult to standardise a test method with issues of

- Whether to use natural rain or artificial watering
- When during a storm to measure the property

- How to achieve consistent water film thickness with natural rain
- How to apply sufficient water at a consistent rate for artificial watering (small amounts of water may drain away into open surfaces if applied too soon before the measurement whilst it may be appropriate for denser surfaces)

As yet, no performance laboratory or *in situ* test has been developed for assessing splash and spray potential that is widely accepted. The techniques most commonly used to measure splash and spray (Sanders et al., 2012) include

- Collection, in which a proportion of the generated splash and spray is collected in a container to provide a representative sample of the splash and spray generated.
- Contrast change, in which images of a standardised target before and during spray are analysed using image analysis technology and the differences used to estimate the amount of spray.
- Light attenuation, in which a light source is directed through a spray cloud at a photocell a fixed distance away, with the light becoming scattered as it passes through the spray, and the amount of light collected by the photocell gives an indication of the quantity of spray.
- Subjective observation, in which either images or direct observation of the spray is assessed by a number of people with each image or test run being scored to produce a subjective quantity of spray.

All these approaches are for *in situ* rather than laboratory tests, so are more difficult for use in asphalt design.

4.4 COLOUR

The colour of the asphalt is an architectural rather than engineering property of a pavement, but it can be useful for aesthetic appeal or, more importantly for highways, to distinguish between different users. One problem with distinguishing lanes for buses, bicycles, pedestrians and/or other categories of traffic is that there is no agreement about which colour should be used for each category. Therefore, the colour scheme used by one road authority may be completely different to those used by adjacent authorities, making it difficult for drivers to be sure what a coloured lane may indicate. It is also claimed that some colours are unacceptable in some areas of cities which have two football teams because they are associated with the 'other' team!

Asphalt is basically black from the colour of bitumen that should coat all the aggregate particles. The 'shade' of black can be influenced by the colour of the aggregate, but the range of colours available is limited and generally relatively muted. Furthermore, the few really 'coloured' natural aggregates do not necessarily have the mechanical properties required in surface course aggregates. Those colours can be enhanced by using clear artificial binders rather than bitumen, although such binders are generally more expensive than bitumen. Therefore, pigments are generally added to the mixture to provide the required coloured, replacing some of the filler, or else a surface treatment is specifically applied to provide the colour.

Another problem with colour is that it will change with time, if only from the detritus masking it. Pigments tend to fade in sunlight and the coloured binder can be abraded from the surface of aggregate particles. When replacing a patch in a coloured surface, it is very difficult to get a reasonable match both when applied and after the patch has also aged.

If colours are to be set, first they need to be defined. The obvious method is to have a colour chart to compare, but these can be extensive and a selection of shades will have to be permitted to allow for the potential variation, both initially and with time. Therefore, there are systems for defining colour using coordinates with a tolerance, with some of the more common systems being listed in Table 4.3. Examples of common colours in three of these codes are given in Table 4.4. There are specialist pieces of equipment to measure these coordinates for each system, some being common.

If colour is to be specified, there needs to be an ageing simulation procedure to replicate the abrasion and exposure that the surface will

Table 4.3 Colour description systems

System	Description
Hex colour code	Hexadecimal triplets representing the colours red, green, and blue as #RRGGBB
RGB colour code	RGB stands for red, green and blue
NCS colour space	Using the three properties of hue, blackness and chromaticness, the latter two being called nuance
HSL colour code	Cylindrical system using hue, saturation and lightness
HSV colour code	Cylindrical system using hue, saturation and value or brightness

Table 4.4 Codes for common colours

Colour	Hex code	RGB code	HSV code
White	#FFFFFF	(255, 255, 255)	(−°, 0%, 100%)
Black	#000000	(0, 0, 0)	(−°, −%, 0%)
Green	#009F6B	(0, 159, 107)	(160°, 100%, 63%)
Red	#C40233	(196, 2, 51)	(345°, 99%, 77%)
Yellow	#FFD300	(255, 211, 0)	(50°, 100%, 100%)
Blue	#0087BD	(0, 135, 189)	(197°, 100%, 74%)

endure in its service life. There are no suitable procedures that are generally accepted, but there is a scheme in the United Kingdom for coloured surface treatments that does have one (BBA, 2005).

REFERENCES

ASTM International. 2010. Standard test method for measurement of hydraulic conductivity of saturated porous materials using a flexible wall permeameter. *ASTM D5084-10*. West Conshohocken, PA: ASTM International.

ASTM International. 2015b. Standard test method for measuring pavement macrotexture depth using a volumetric technique. *ASTM E965-15*. West Conshohocken, PA: ASTM International.

ASTM International. 2015c. Standard practice for calculating pavement macrotexture mean profile depth. *ASTM E1845-15*. West Conshohocken, PA: ASTM International.

Austroads. 2011. Pavement surface texture measurement with a laser profilometer. *Test Method AG:AM/T013*. Sydney: Austroads. http://austroads. com.au.tmp. anchor.net.au/images/stories/AG_AM_T013__Texture_survey__2011.pdf

British Board of Agrément. 2005. *Guidelines Document for the Assessment and Certification of Coloured Surface Treatments for Highways*. Watford: British Board of Agrément. www.bbacerts.co.uk/wp-content/uploads/2014/10/Coloured-Surface-Treatments-For-Highways-Guidelines.pdf

Comité Européen de Normalisation. 2002a. Aggregates for bituminous mixtures and surface treatments for roads, airfields and other trafficked areas. *EN 13043:2002*. London: BSI; Berlin: DIN; Paris: AFNOR; and other European standards institutions.

Comité Européen de Normalisation. 2002b. Road and airfield surface characteristics – Test methods – Part 3: Measurement of pavement surface horizontal drainability. *EN 13036-3:2002*. London: BSI; Berlin: DIN; Paris: AFNOR; and other European standards institutions.

Comité Européen de Normalisation. 2003b. Road and airfield surface characteristics – Test methods – Part 7: Irregularity measurement of pavement courses – The straightedge test. *EN 13036-7:2003*. London: BSI; Berlin: DIN; Paris: AFNOR; and other European standards institutions.

Comité Européen de Normalisation. 2008b. Road and airfield Surface characteristics – Test methods – Part 8: Determination of transverse unevenness indices. *EN 13036-8:2008*. London: BSI; Berlin: DIN; Paris: AFNOR; and other European standards institutions.

Comité Européen de Normalisation. 2009b. Tests for mechanical and physical properties of aggregates – Part 8: Determination of the polished stone value. *EN 1097-8:2009*. London: BSI; Berlin: DIN; Paris: AFNOR; and other European standards institutions.

Comité Européen de Normalisation. 2010a. Road and airfield surface characteristics – Test methods – Part 1: Measurement of pavement macrotexture depth using a volumetric patch technique. *EN 13036-1:2010*. London: BSI; Berlin: DIN; Paris: AFNOR; and other European standards institutions.

Comité Européen de Normalisation. 2011. Road and airfield surface characteristics – Test methods – Part 4: Method for measurement of slip/skid resistance of a surface – The pendulum test. *EN 13036-4:2011.* London: BSI; Berlin: DIN; Paris: AFNOR; and other European standards institutions.

Comité Européen de Normalisation. 2012g. Bituminous mixtures – Test methods for hot mix asphalt – Part 40: In situ drainability. *EN 12697-40:2012.* London: BSI; Berlin: DIN; Paris: AFNOR; and other European standards institutions.

Comité Européen de Normalisation. 2014b. Bituminous mixtures – Test methods for hot mix asphalt – Part 49: Determination of friction after polishing. *EN 12697-49:2014.* London: BSI; Berlin: DIN; Paris: AFNOR; and other European standards institutions.

Daskova, J and J Kudrna. 2012. The experience with Wehner/Schulze procedure in the Czech Republic. *Procedia – Social and Behavioral Sciences.* In SIIV – 5th International Congress – Sustainability of Road Infrastructures 2012, Volume 53, 3 October 2012, pp 1034–1043. http://www.sciencedirect.com/science/article/pii/S1877042812044151

Dunford, A and P G Roe. 2012a. *Use of the Wehner–Schulze machine to explore better use of aggregates with low polish resistance – 1: Capabilities of the Wehner–Schulze machine.* TRL Published Project Report PPR604. Wokingham: TRL Limited.

Dunford, A and P G Roe. 2012b. *Use of the Wehner–Schulze machine to explore better use of aggregates with low polish resistance – 2: Experiments using the Wehner–Schulze machine.* TRL Published Project Report PPR605. Wokingham: TRL Limited.

Friel, S, M Kane and D Woodward. 2013. Use of Wehner Schulze to predict skid resistance of Irish surfacing materials. In Airfield and Highway Pavement, June 2013, France, 12 p. https://hal.archives-ouvertes.fr/hal-00851551/document

International Organization for Standardization. 2001. Acoustics – Measurement of the influence of road surfaces on traffic noise – Part 1: Statistical Pass-By method. *ISO 11819-1:2001.* International standards institutions.

International Organization for Standardization. 2004. Characterization of pavement texture by use of surface profiles – Part 1: Determination of mean profile depth. *ISO 13473-1:2004.* International standards institutions.

International Organization for Standardization. 2013. Acoustics – Measurement of the influence of road surfaces on traffic noise – Part 4: SPB method using backing board. *ISO/PAS 11819-4:2013.* International standards institutions.

Nicholls, J C. 1997. *Review of UK porous asphalt trials.* TRL Report TRL264. Wokingham: TRL Limited.

Nicholls, J C and I G Carswell. 2001. *Effectiveness of edge drainage details for use with porous asphalt.* TRL Report TRL376. Wokingham: TRL Limited.

Nicholls, J C and M E Daines. 1992. Spray suppression by porous asphalt. In *The Second International Symposium on Road Surface Characteristics*, Berlin, 1992.

Nicholls, J C, C Roberts and P Samuel. 2006a. *Implications of implementing the European asphalt test methods.* TRL Report TRL656. Wokingham: TRL Limited.

Roe, P G and S A Hartshorne. 1998. *The polished stone value of aggregates and in-service skidding resistance.* TRL Report TRL322. Wokingham: TRL Limited.

Roe, P G, A R Parry and H E Viner. 1998. *High- and low-speed skidding resistance: The influence of texture depth.* TRL Report TRL367. Wokingham: TRL Limited.

Roe, P G, D C Webster and G West. 1991. *The relation between the surface texture of roads and accidents.* TRL Research Report RR296. Wokingham: TRL Limited.

Sanders, P D, A Dunford, H Viner, G W Flintsch and R M Larson. 2012. *Splash and spray assessment tool development program – First interim report: Revised synthesis report.* TRL Published Project Report PPR602. Wokingham: TRL Limited.

The Highways Agency, Transport Scotland, Welsh Assembly Government and The Department for Regional Development, Northern Ireland. 2006a. Surfacing materials for new and maintenance construction. In *Design Manual for Roads and Bridges: Volume 7, Pavement Design and Maintenance: Section 5, Surfacing and Surfacing Materials: Part 1, HD 36/06.* London: The Stationery Office. www.standardsforhighways.co.uk/dmrb/vol7/section5/hd3606.pdf

The Highways Agency, Transport Scotland, Welsh Assembly Government and The Department for Regional Development, Northern Ireland. 2008a. Specification for Highway Works, Series 900, Road Pavements – Bituminous Bound Materials. In *Manual of Contract Documents for Highway Works, Volume 1.* London: The Stationery Office. www.standardsforhighways.co.uk/ha/standards/mchw/vol1/pdfs/series_0900.pdf

The Highways Agency, Transport Scotland, Welsh Assembly Government and The Department for Regional Development, Northern Ireland. 2008b. Notes for guidance on the specification for highway works, series NG 900, road pavements – Bituminous bound materials. In *Manual of Contract Documents for Highway Works, Volume 2.* London: The Stationery Office. www.standardsforhighways.co.uk/ha/standards/mchw/vol2/pdfs/series_ng_0900.pdf

Viner, H, P Abbott, A Dunford, N Dhillon, L Parsley and C Read. 2006. *Surface texture measurement on local roads.* TRL Published Project Report PPR148. Wokingham: TRL.

Chapter 5

Surface course properties

5.1 GENERAL

The surface course properties discussed in this chapter are in addition to the surface characteristics discussed in Chapter 4.

5.2 DEFORMATION RESISTANCE

5.2.1 Issue being addressed

An uneven surface creates problems with drainage when wet, with water collecting in low spots during periods of wet weather that can lead to both increased spray (resulting in reduced driver visibility) and reduced tyre adhesion to the pavement surface. In extreme cases, aquaplaning can occur with complete loss of grip on the surface. Deformation can also create difficulties for drivers to follow their intended path, particularly with deep ruts from which it can be difficult to leave when diverging from the path taken by the majority of heavier vehicles on that section of the pavement.

Pavements are generally built with a smooth surface profile (Section 4.1.2) but can develop ruts from the loads applied by vehicles. The amount of deformation depends on a number of factors, in particular:

- The load applied by vehicle wheels, with an empirical relationship of the deformation impact being to the fourth power of the load applied, so that heavy vehicles with few wheels are more damaging than lighter vehicles or those with more wheels.
- The loading frequency, which is greater with both greater traffic flows and more channelling of that traffic. If the wheel loads are evenly distributed across the whole of the pavement, the deformation should be uniform and only be a concern in terms of adjacent structures including the drainage and restraining barriers.
- The ambient temperature is critical because deformation resistance of asphalt is highly dependent on temperature, with the majority

of the deformation occurring on the few warmest days of the year. Therefore, more deformation tends to occur on locations exposed to sunlight, like south-facing slopes in the northern hemisphere, rather than those in the shade.

- Also, new asphalt surfaces tend to be blacker than older ones and, consequently, will absorb more heat; this effect is unfortunate because new asphalt is also more susceptible to deformation because it hardens with age.
- The loading period, with the amount of deformation increasing with slower traffic speeds.
- The resistance of the pavement to deformation.

Because it is the only factor that can be controlled, the resistance of the pavement to deformation has to be selected so as to be able to take the traffic loads, loading frequency and expected temperature profile of the site.

The deformation resistance of asphalt generally refers to the secondary compaction of the surface layers, as shown in Figure 5.1, which occurs primarily in the surface course with some in binder course and little in the base because the loads are dispersed as they progress down the pavement. There is also structural deformation, which results from inadequate pavement (rather than asphalt) design when the stiffness of the bound layers is insufficient (Section 6.1). Structural deformation occurs at lower levels and tends to be more extensive because it involves 'bending' the bound pavement; hence being regarded as occurring at the interface with the unbound layers.

Figure 5.1 Cross-section through deformed surfacing. (Courtesy of Dr. Mike E Nunn.)

5.2.2 Conventional measures

Most mixture types will be susceptible to deformation if the air voids content is particularly low. Therefore, there is often a minimum air voids content limit, set at a relatively low value, for deformation as well as a maximum air voids content limit for durability. In a similar manner, the US Superpave has a restricted zone in the grading envelope depending on the nominal maximum aggregate size to restrict deformation as given in AASHTO MP2 (AASHTO, 2004). It has been suggested that the restricted zone requirement is redundant (Kandhal and Cooley, 2001) but others have found the restricted zone does work.

Other conventional measures that can be used to encourage a deformation resistant mixture are the use of harder grade bitumen and/or the use of an aggregate grading relying on aggregate interlock for strength rather than mortar, which is highly temperature susceptible. This susceptibility is shown by the results of wheel-tracking tests, with the values rising noticeably faster with increasing temperature for hot rolled asphalt (which is a mortar-based mixture) than for stone mastic asphalt or asphalt concrete (which are aggregate interlock mixtures). Of these two options, the binder grade is the more commonly adopted measure because the choice of aggregate grading is dictated by other considerations.

5.2.3 Marshall stability (simulative)

The Marshall tests for stability and flow, together with the volumetrics, were developed as a design method for asphalt mixtures, particularly the optimum binder content for a particular aggregate grading. However, the stability characteristic has been, and still is, used as a measure of the deformation resistance of the mixture. An example of its use is that the test to EN 12697-34 (CEN, 2012h) is the parameter for use on airfields in the European standards while the American version of test is ASTM D6927 (ASTM, 2015d).

The specimens for the test are cylinders that are compacted by an impact compactor for a set number of blows at a set temperature. The number of blows is usually 50 per side but, for example, 35 blows may be used for light traffic or 75 for heavy duty pavements. The compacted specimens are de-moulded and allowed to cool in air before being put in a water bath at the test temperature, usually 60°C. The test specimens are then put into a breaking head (Figure 5.2) and the upper and lower breaking heads are moved relative to each other at a constant rate of deformation. The maximum load applied is the stability.

The deformation resistance is assumed to increase with greater values of Marshall stability. The categories of Marshall stability offered in EN 13108-1 (CEN, 2016a) for asphalt concrete on airfields in Europe have minima of 2.5 kN up to 12.5 kN rising by increments of 2.5 kN plus a no requirement category. The precision given is a repeatability of 1.7 kN and

Figure 5.2 Breaking head mould for Marshall test. (Courtesy of John Prime.)

reproducibility of 2.2 kN for Marshall stability a repeatability of 0.7 mm and reproducibility of 0.8 mm for flow.

5.2.4 Wheel tracking (simulative)

The wheel-tracking test is the most obvious simulative test for deformation resistance by running a wheel repeatedly over a slab of the asphalt mixture. However, there are a number of aspects that have to be standardised including

- The diameter and width of the wheel
- The type of tyre (pneumatic, solid rubber or steel)
- The load on the wheel
- The frequency of loading
- The total number of load applications
- The test temperature
- The test result (rut depth, rut depth as a proportion of slab depth or tracking rate towards the end of the test)

In addition, the test can be carried out in air, when the result is purely a measure of deformation resistance, or in water, when the result is a measure of water sensitivity as well as deformation resistance. The test temperature can be controlled more precisely in water than air, although adequate control can generally be achieved with air.

In the test, the samples for the wheel-tracking test are generally slab compacted, although some methods can be undertaken on cores recovered from site. Either the wheel is travelled back and forward over the slab or the slab is moved back and forward under the wheel at a controlled temperature and frequency for a set number of passes.

Figure 5.3 Large device for CEN wheel tracking. (Courtesy of John Prime.)

In the European test EN 12697-22 (CEN, 2003c), the options were not able to be totally harmonised, with three different devices (the extra-large device with pneumatic tyres, the large device with pneumatic tyres as in Figure 5.3 and the small device with solid rubber tyres as in Figure 5.4), although only the large and small devices are accepted for CE Marking. Furthermore, there are two procedures with the small device and one of them permits water conditioning as well as the air conditioning that is common to the procedures with other two devices.

In America, the Hamburg wheel-track test to AASHTO T-324 (AASHTO, 2014) is used. This test is carried out with a steel wheel in water and is primarily used to assess water sensitivity rather than deformation resistance.

Figure 5.4 Small device for CEN wheel tracking. (Courtesy of John Prime.)

Because of the number of options, several wheel tracking tests have been developed with significant differences. The results from the different test methods have been found not to correlate well with each other (Bonnot, 1997; Nicholls et al., 2006a), demonstrating that these selections are important. Whilst easily deformed mixtures generally perform badly and stiffer mixtures generally perform well, the precise ranking of different mixtures will differ depending on the options selected.

Part of reason for the different results is the use of different measures to determine the deformation, in particularly the wheel-tracking rate over the last part of the test, the total rut depth and the total rut depth as a proportion of the sample thickness. Using the rut depth assumes that the sample is sufficiently thick for the wheel-load to be sufficiently distributed at the bottom of the sample for no further deformation to occur there, while using the proportional rut depth assumes that the whole depth of the sample is deformed equally. In practice, the truth is somewhere between the two assumptions but the implication tends to be small provided the samples are of roughly similar depths.

The lack of a definitive wheel-tracking test is also demonstrated by a method of predicting the permanent deformation of HRA (Szatkowski and Jacobs, 1977) that was further developed in 1998. The 'best' equation found to predict the development of permanent deformation in a HRA surface course from the wheel-tracking rate and other factors had a correlation coefficient, R_{adj}^2, of only 0.72. However, there were inconsistencies when the data were extrapolated to other potential situations including when trying to incorporate ambient temperature as a parameter in a logical manner. Unfortunately, the latter work was never published.

The deformation resistance is assumed to increase with rut depth and reduced rate of rutting. The categories for specifying deformation resistance by the wheel-tracking test that EN 13108 gives for Europe are listed in Table 5.1.

The range of minima value categories for the different test measures demonstrates that the value of any criterion is linked to a particular test method and that a different value should be used if any aspect of the test changes to maintain the same level of performance in terms of expected *in situ* rutting. The amount that the criterion needs to be adjusted depends on the extent to which the aspect affects the result. This variation also applies to the limits set for all other properties.

The precision given in EN 12697-22 for the proportional rut depth with the large-sized device are

- A repeatability of 0.76% and reproducibility of 0.97% after 100 passes; rising to
- A repeatability of 1.12% and reproducibility of 1.16% after 30,000 passes.

Table 5.1 European categories for wheel tracking in EN 13108

Device	Large size	Small size				
		Procedure A		Procedure B		
Measure	Proportional rut depth (%)	Rate (µm/cycle)	Rut depth (mm)	Rate (µm/cycle)	Proportional rut depth (%)	Rut depth (mm)
Categories	<5.0	<5.0	<3.0	<0.02[a]	<1.0[b]	<1.0
				<0.03	<1.5[b]	<1.5
		<7.5	<5.0	<0.04	<2.0[b]	<2.0
	<7.5			<0.05	<2.5[b]	<2.5
		<10.0	<7.0	<0.06	<3.0	<3.0
				<0.07	<4.0	<3.5
	<10.0	<12.5	<9.0	<0.08[c]	<5.0	<4.0
				<0.09[c]	<6.0	<4.5
		<15.0	<11.0	<0.10	<7.0	<5.0
	<15.0			<0.15	<8.0	<6.0
		<17.5	<13.0	<0.30	<9.0	<6.5
				<0.40	<11.0	<7.0
	<20.0	<20.0	<16.0	<0.50	<13.0	<8.0
				<0.60	<16.0	<9.0[d]
	–	–	–	<0.80	<19.0[b]	<10.0[d]
	–	–	–	<1.00	<20.0[d]	–
	–	–	–	–	<25.0[d]	–
			No requirement			
Asphalt types	AC, BBTM, SMA, PA	HRA	HRA	AC, SMA, PA	AC, SMA, PA	AC, SMA, PA

[a] PA only.
[b] SMA only.
[c] Not for PA.
[d] AC only.

The precision for the wheel-tracking rate to Procedure A with the small-sized device are given as

- A repeatability of 0.5 µm/cycle and a repeatability of 1.0 µm/cycle for laboratory-prepared samples at a level of 2.1 µm/cycle
- A repeatability of 0.6 µm/cycle and a repeatability of 1.1 µm/cycle for laboratory-prepared samples at a level of 1.7 µm/cycle
- A repeatability of 2.5 µm/cycle and a repeatability of 4.7 µm/cycle for cored samples at a level of 6.4 µm/cycle
- A repeatability of 3.2 µm/cycle and a repeatability of 4.5 µm/cycle for cored samples at a level of 10.7 µm/cycle

5.2.5 Indentation test (simulative)

Mastic asphalt is a mortar-based mixture that has relative small maximum nominal aggregate size that is more susceptible to indentation from static loads rather than rutting from moving tyres. Therefore, there are specialist tests for indentation of mastic asphalt mixtures. In Europe, there are two such tests, EN 12697-20 (CEN, 2012i) using cube or cylindrical specimens and EN 12697-21 (CEN, 2012j) using plate (slab) specimens. Both tests are limited to aggregate sizes not greater than 16 mm but can be used for such mixtures other than mastic asphalt.

The tests determine the depth of indentation into the asphalt when a force is applied via a cylindrical indentor pin with a circular flat-ended base for a fixed time at a standard temperature. The result is the arithmetic mean of a number of separate determinations, and is called the hardness number in EN 12697-21.

The lesser the indentation or hardness number, the greater the resistance to indentation. EN 13108-6 (CEN, 2016f) gives a series of categories for indentation as given in Table 5.2. There are also no requirement categories for each property.

The precision of EN 12697-20 are a repeatability of 28% and reproducibility of 55% while that of EN 12697-21 are repeatabilities of 42% and 28% and reproducibilities of 69% and 56% for hardness numbers less or more than 40×0.1 mm, respectively.

5.2.6 Cyclic compression (fundamental)

The cyclic compression test is widely considered to be the fundamental test for deformation resistance, although it is really just a more sophisticated simulative test. Originally, the test was on unconfined samples, when it was a simulative test, but variants have been developed with increasing sophistication including uniaxial and triaxial compression tests of which the triaxial variant is nominally the fundamental test. The European test, EN 12697-25 (CEN, 2016j), contains both options.

Table 5.2 Categories for resistance to indentation

Property	Maximum category	Minimum category
Minimum indentation (mm)	1.0	3.0
Maximum indentation (mm)	2.5	15.0
Maximum increase of indentation after 30 min (mm)	0.3	0.8
Maximum cumulative deformation after 2500 load cycles (mm)	1.0	4.5
Maximum cumulative deformation after 5000 load cycles (mm)	1.0	4.5

The uniaxial variant can be undertaken on cylindrical test specimens prepared in the laboratory or cored from a pavement. A specimen is maintained at an elevated conditioning temperature and placed between two parallel loading platens of which the upper platen has a diameter that is significantly less than that of the specimen. The specimen is subjected to a cyclic axial block-pulse pressure with no additional lateral confinement pressure. The change in height of the specimen is measured at specified numbers of load applications and the cumulative axial strain is determined to form a creep curve from which the creep characteristics can be identified.

The triaxial variant can also be undertaken on cylindrical test specimens prepared in the laboratory or cored from a pavement. A specimen at an elevated conditioning temperature is placed between two plane parallel loading platens of the same diameter as the specimen. The specimen is subjected to a confining pressure, which can be either static or dynamic, with a cyclic axial pressure also being superimposed that can be either haversinusoidal, to which rest periods can be applied, or block-pulse. In both cases, a small axial dead load may be applied. Again the change in height of the specimen is measured at specified numbers of load applications from which the cumulative axial strain is determined to form a creep curve from which the creep characteristics are identified.

The deformation resistance is assumed to increase with reduced creep rates. EN 13108 gives categories of triaxial creep rate that can be specified in Europe for asphalt concrete, stone mastic asphalt and porous asphalt with maxima of 0.2 µm/m/n to 1.6 µm/m/n in increments of 0.2 µm/m/n and then 2.0 µm/m/n to 16.0 µm/m/n in increments of 2.0 µm/m/n plus a no requirement category. The precision given in EN 12697-25 are

- A repeatability of 17.3% and reproducibility of 21.5% for the uniaxial test
- A repeatability of 0.8% and a reproducibility of 1.6% for the triaxial test with AC
- A repeatability of 0.41% and a reproducibility of 0.73% for the triaxial test with HRA
- A repeatability of 0.17% and a reproducibility of 0.17% for the triaxial test with SMA

5.2.7 Specifying deformation resistance

The extent to which deformation resistance needs to be specified will be dependent on the traffic loads expected to be applied to the pavement and the importance of avoid rutting on it. On roads with limited traffic and no great significance, there is no reason not to rely on the conventional measures to control the property. However, the fundamental and simulative methods at reduced category levels are needed with greater traffic levels. The fundamental cyclic compression test should be the preferred option if its fundamental status is accepted, but the wheel-tracking tests do 'feel' more realistic. Of the options within the cyclic compression test, the precision data do support the preference for the triaxial method that is called up in EN 13108. The indentation test is appropriate for mastic asphalt and other asphalts with small aggregate sizes when indentation from point loads is anticipated. The use of Marshall stability is not recommended because it is not specifically designed for deformation and, if testing is required, it might as well be a test designed for deformation. However, many airfield owners still require the use of Marshall stability requirements.

5.3 RESISTANCE TO CRACKING

5.3.1 Issue being addressed

There are a many types and causes of cracking that can occur in asphalt pavements that, initially at least, may not be detrimental to the pavement. However, cracks can lead to water being able to get into the pavement more easily and/or the loss of aggregate particles as the cracks widen, both of which will result in a loss of performance and may result in the structural failure of the pavement. Water residing within a pavement can lead to stripping of the asphalt from the aggregate (Section 7.2) while loss of aggregate can result in ravelling and/or potholes (Section 5.4.3).

The types of cracking in asphalt pavement include alligator, block (Figure 5.5), longitudinal (Figure 5.6) and reflective cracks while the causes include fatigue (Section 6.2.3), temperature fluctuations and differential movement at lower layers. Joints in the mat, while often being necessary for practical reasons, can often initiate or exacerbate cracking problems.

Figure 5.5 Example of block cracking.

Figure 5.6 Example of longitudinal cracking. (Courtesy of Nigel Hewitt.)

Because of the variety of types and causes, no fundamental mechanism has been identified and, therefore, no fundamental test has been developed.

5.3.2 Conventional measures

Conventional measures have not generally been explicitly applied for crack resistance, but the presence of sufficient mortar (which implies high contents of binder, filler and fine aggregate) plus the use of binders with good elasticity, possibly enhanced by polymer modification, are generally expected to improve the property. Furthermore, as with many other required properties of asphalt, adequate compaction on site will also be required to ensure good performance.

5.3.3 Fatigue cracking

Fatigue cracking results from fatigue damage of the structural layers that causes the cracking in all the layers above because they are not adequately supported. The tests for fatigue resistance are covered in Section 6.2.

5.3.4 Tensile strength (simulative)

Neither compressive nor tensile strength properties are used in most procedures for the design of the asphalt pavements, the stiffness modulus being used instead. Furthermore, asphalt is stronger in compression than tension. However, the tensile strength is often required as a surrogate for crack resistance because it is the tensile strength that will resist the initiation and propagation of cracks.

The tensile stiffness test is a destructive test in which a cylindrical specimen at a specific test temperature is placed in a compression testing machine between loading strips and loaded diametrically along the direction of the cylinder axis with a constant speed of displacement until it breaks. The indirect tensile strength is the maximum tensile stress at the peak load applied when the break occurs. The European test procedure is EN 12697-23 (CEN, 2003d), while the American version is ASTM D6931 (ASTM, 2012a).

Higher tensile stiffness values should indicate greater resistance to cracking, among other properties. However, no specification categories are given in EN 13108 for tensile stiffness while EN 12697-23 states that precision data for testing at 5°C and for testing on cylindrical specimens have not yet been established.

5.3.5 Low-temperature cracking (simulative)

Tensile stresses build up in a pavement as the temperature falls because the asphalt is, to a greater or lesser extent, restrained and, if those stresses reach a critical level, cracks will form. Such thermal cracks can be initiated by one particularly low temperature or by multiple warming and cooling cycles, with the latter being exacerbated by fatigue from traffic loadings. Low-temperature cracking is the most prevalent distress found in asphalt pavements built in cold weather climates.

The European test method, EN 12697-46 (CEN, 2012k), has five different test methods for the low-temperature performance of asphalt specimens. These procedures are

- Uniaxial tension stress test (UTST) in which a specimen is pulled with constant strain at a constant temperature until failure occurs to determine the maximum stress and the corresponding tensile failure strain (Figure 5.7a).
- Thermal stress restrained specimen test (TSRST) in which a specimen is held at constant length while the temperature decreases uniformly, building up cryogenic stress in the specimen, to determine the stress and temperature at failure (Figure 5.7b).
- Tensile creep test (TCT) in which the specimen is subjected to a constant tension stress at a constant temperature for a time before the stress is withdrawn with the strain being monitored so that rheological

parameters describing the elastic and viscous properties of the asphalt can be determined (Figure 5.7c).

- Relaxation test (RT) in which the specimen is subjected to a spontaneous strain that is held constant with the tension stress, reducing with time by relaxation, being monitored to determine the time of relaxation and the tension stress remaining at the end of the test (Figure 5.7d).
- Uniaxial cyclic tension stress test (UCTST) in which a specimen is subjected to a cyclic sinusoidal tensile stress to simulate dynamic traffic loading together with a constant stress to simulate cryogenic stress. The stress and strain are monitored until fatigue failure occurs

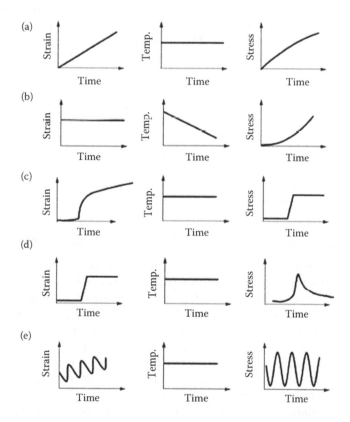

Figure 5.7 Principle of European low-temperature crack tests. (a) Uniaxial tension stress test. (b) Thermal stress restrained specimen test. (c) Tensile creep test. (d) Relaxation test. (e) Uniaxial cyclic tension stress test. (Adapted from Spiegl, M. 2007. Tieftemperaturverhalten von bitumiösen baustoffen – Labortechnische ansprache und numerische simulation des gebrauchsverhaltens (Low temperature behaviour of bituminous construction materials – Assessment on laboratory scale and numerical simulation of the low temperature performance). Doctoral thesis. Techische Universtität Wien.)

to determine the number of load cycles to failure plus the number of load cycles until the conventional fatigue criterion is reached (Figure 5.7e).

The variety of these tests demonstrates the number of different aspects that can be measured and the lack of agreement as to which needs to be controlled in order to minimise low-temperature cracking. However, it is the maximum failure temperature and associated starting temperature and temperature rate from the TSRST that is planned to be used for CE Marking.

Low failure temperatures from the TSRST option indicate that low-temperature cracking should be reduced. The categories given in EN 13108 for all mixture types except soft asphalt are maxima failure temperature of −15°C to −30°C in 2.5°C increments (and −12.5°C for mastic asphalt only) plus a no requirement category. The omission of soft asphalt is interesting because soft asphalt is predominately used in Scandinavia, where low temperatures are common. EN 12697-46 reports that no precision data are available.

The American Superpave specification approaches low-temperature cracking by specifying the properties of the binder using ASTM D6816 (ASTM, 2011). Testing the binder makes it a conventional measure because it is on a component material rather than on the mixture. However, it has been found that the asphalt mixture itself needs to be tested before low-temperature cracking properties can be accurately predicted (Minnesota Department of Transportation, 2016).

5.3.6 Resistance to crack development (simulative)

The semi-circular bending (SCB) test method determines the tensile strength or fracture toughness of an asphalt mixture which is regarded as a surrogate for the potential for crack propagation. Crack propagation is the second stage of crack development following crack initiation, with different mixtures not necessarily performing similarly for both aspects. Crack initiation is basically covered by fatigue cracking (Section 5.3.3).

In the European test, EN 12697-44 (CEN, 2010b), a cylinder of asphalt, often a core, is cut across the diameter with a thin central crack cut transversely across the test sample. This sample is loaded in three-point bending so that the middle of the base is subjected to a tensile stress. The resulting deformation is increased at a constant rate to a maximum value that is directly related to the fracture toughness, defined as resistance to failure of the test piece by breaking. The results are used to calculate the maximum load that the material containing a notch (crack) can resist before failure and when the presence of a notch significantly affects the maximum load reached.

High fracture toughness is expected to lead to reduced crack propagation. The categories put forward for specifying the fracture toughness in

EN 13108 for asphalt concrete and SMA have minima fracture toughness of 10 N/mm$^{1.5}$ to 55 N/mm$^{1.5}$ in increments of 5 N/mm$^{1.5}$. The precision for the SCB test stated in EN 12697-44 is a repeatability of 2.44 N/mm$^{1.5}$ and a reproducibility of 2.49 N/mm$^{1.5}$.

The equivalent American test is ASTM E647 (ASTM, 2015e).

5.3.7 Specifying resistance to cracking

The problem with specifying for resistance to cracking is that there are several different forms of cracking. However, all the measures will give some indications of the resistance to cracking, even when the test is intended to measure a different form of cracking. In general, the conventional measures should be adequate to resist cracking on a well-designed flexible pavement. When there is any obvious potential for cracking, such as reflective cracks or low-temperature cracks, then the mixture should be checked against the relevant test.

5.4 RESISTANCE TO LOSS OF AGGREGATE

5.4.1 Issue being addressed

Loss of aggregate particles is one of the most common failure modes of asphalt pavements. It is the progressive dislodgement and loss of fine and then coarse aggregate from the road surface (Nikolaides, 2015) by the passage of traffic or weathering. Once some particles have been lost, the deterioration becomes progressively faster with time, with the various stages usually being described as light (loss of surface fines), moderate (loss of fines and some coarse aggregate particles) and severe (loss of fine and coarse aggregate).

The loss occurs when the micro-mechanical bond between binder and aggregate reaches a critical point. Cohesive fracture of the binder will occur when the tensile stress (induced in the binder as a result of the movement) exceeds the breaking stress of the binder and that fracturing will result in the aggregate particles being detached from the pavement surface. Thus, loss of aggregate is most likely to occur at low temperatures and at short loading times when the stiffness of the binder is high (Hunter et al., 2015).

The effect of losing aggregate is that the surface has an open appearance with loose aggregate on and around the pavement and remaining aggregate particles being easy to remove from the surface. As the loss progresses, the surface becomes rough and 'pock marked', with those 'pock marks' developing into potholes if left untreated (Thom, 2014; Nikolaides, 2015). Loss of aggregate also results in surface roughness/unevenness, water collecting in the depressions left, reduced skid resistance, less comfort for drivers and more noise.

The loss of aggregate from an asphalt pavement surface is called by several different names. Fretting is the loss of the mortar and fine aggregate, which removes the support for larger aggregate particles and hence their loss. Ravelling is the plucking out of coarse aggregate particles, leaving the mortar exposed to abrasion from passing vehicle tyres. The order may differ, but the final result is the same. Scuffing is another term that has been used as an acronym for ravelling or fretting, but the term can be confused with tyre scuffing, where tyre rubber is left on asphalt making it look unsightly. Tyre scuffing will not affect the integrity or durability of the pavement surface whereas ravelling will adversely affect performance and durability.

5.4.2 Conventional measures

Conventional measures have not generally been explicitly applied for fragmentation, but the same properties as for crack resistance (Section 5.3.2) of sufficient mortar, the use of binders with good elasticity and adequate compaction on site have been used implicitly.

5.4.3 Resistance to scuffing (simulative)

The importance of the failure mechanism has not been matched by the development of a test to measure the potential to resist it. However, a test method is being developed at the time of writing that should be published as European Technical Specification CEN/TS 12697-50 in due course. The draft test method is intended to determine the resistance to scuffing of asphalt mixtures which are used in surface layers and are loaded with high shear stresses in road or airfield pavement.

The test uses either laboratory-compacted asphalt specimens or specimens cut from a pavement. The specimen is fixed in a test facility and loaded simultaneously with both normal and shear stresses. Due to these stresses, material loss will occur from the surface of the slab. This material loss depends on the resistance to scuffing of the tested asphalt mixture: the higher the resistance, the less material will be dislodged.

However, the current draft has four different kinds of loading facilities with which to undertake the test:

- The Aachener ravelling tester (ARTe, Figure 5.8)
- The Darmstadt scuffing device (DSD, Figure 5.9)
- The rotating surface abrasion test (RSAT, Figure 5.10)
- The Triboroute (Figure 5.11)

In order to try to rationalise the four devices, the Conferences of European Directors of Roads (CEDR) has commissioned a research project to obtain comparative results and this project is underway at the time of writing.

Figure 5.8 Aachener ravelling tester. (Courtesy of BAM Infra Asfalt.)

Figure 5.9 Darmstadt scuffing device. (Courtesy of Centre de Recherches Routières.)

5.4.4 Particle loss (simulative)

Porous asphalt or open-graded friction course mixtures are more susceptible to aggregate particle loss and, therefore, a specific test has been developed for such mixtures with the European method being EN 12697-17 (CEN, 2004) and the American method being Appendix X2 of ASTM D7064/ D7064M (ASTM, 2013a). The test is also widely known as the Cantabro test after the university where it was originally developed.

Particle loss is assessed by the loss of mass of laboratory-prepared cylindrical samples of porous asphalt after they have been rotated in a Los Angeles machine in order to estimate the abrasiveness of porous asphalt – not many

Figure 5.10 Rotating surface abrasion test device. (Courtesy of Heijmans Infrastructuur en Milieu.)

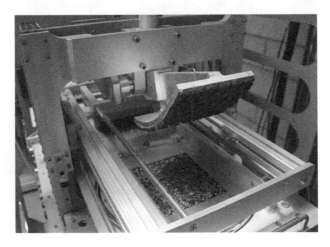

Figure 5.11 Triboroute device. (Courtesy of Institut Français des Sciences et Technologies des Transports, de l'Aménagement et des Réseaux.)

pavements are stressed as extremely as this test imposes. The test does not reflect the abrasive effect by studded tyres.

The less the particle loss result is, the more robust the asphalt mixture is assumed to be. EN 13108-7 (CEN, 2016b) gives categories for use with porous asphalt in Europe with maxima particle loss of 10%, 15%, 20%, 30%, 40% and 50% plus a no requirement category. The precision given in EN 12697-17 for particle loss and repeatabilities of 2% and 5% and reproducibilities of 4% and 8% for levels of up to and over 40%, respectively.

5.4.5 Abrasion by studded tyres (simulative)

Studded tyres are not used, or even permitted, in most countries, but the very abrasion action of such tyres does need to be considered in areas where studded tyres are used to counter frozen conditions, such as in most Scandinavian countries. European standard EN 12697-16 (CEN, 2016k) includes two methods for determining the susceptibility of abrasion by studded tyres using cylindrical samples of the mixtures. The samples can be laboratory-produced or cores drilled from a slab or pavement and the aggregate has an upper sieve size not exceeding 22 mm.

One method is based on the Prall method, which has been improved by comprehensive Nordic research work. The specimen is brought to a temperature of 5°C and then worn by abrasive action from 40 steel spheres for 15 min. The abrasion value is the loss of volume in millilitres. The results correlate well with abrasion in the field when using paving grade bitumen but the correlation between laboratory and abrasion in field has not been established when polymer- or rubber-modified bitumen is used.

The other method originates from Finnish experience and is also suitable when polymer-modified, but not rubber-modified, bitumen is used. The specimen is again brought to a temperature of 5°C and then worn with water present by three studded tyres for 2 h. The abrasion value is again the loss in millilitres.

5.4.6 Specifying against loss of aggregate

Abrasion by studded tyres and particle loss from porous asphalt are for specific situations and are not necessarily appropriate for general use. However, until there is a fully standardised method for assessing the potential for scuffing, those two tests may need to be used for other situations if conventional measures are considered inadequate.

REFERENCES

American Association of State Highway and Transportation Officials. 2004. Superpave volumetric mix design. *AASHTO MP2*. Washington, DC: AASHTO.

American Association of State Highway and Transportation Officials. 2014. Standard method of test for Hamburg wheel-track testing of compacted hot mix (HMA). *AASHTO T-324*. Washington, DC: AASHTO.

ASTM International. 2011. Standard practice for determining low-temperature performance grade (PG) of asphalt binders. *ASTM D6816-11*. West Conshohocken, PA: ASTM International.

ASTM International. 2012a. Standard test method for effect of heat and air on a moving film of asphalt (rolling thin-film oven test). *ASTM D2872-12e1*. West Conshohocken, PA: ASTM International.

ASTM International. 2013a. Standard practice for open-graded friction course (OGFC) mix design. *ASTM D7064/D7064M-08*. West Conshohocken, PA: ASTM International.

ASTM International. 2015d. Standard test method for Marshall stability and flow of asphalt mixtures. *ASTM D6927-15*. West Conshohocken, PA: ASTM International.

ASTM International. 2015e. Standard test method for measurement of fatigue crack growth rates. *ASTM E647-15*. West Conshohocken, PA: ASTM International.

Bonnot, J. 1997. Prenormative research on wheel tracking test. CEN TC227/WG1 Paper N633. Unpublished.

Comité Européen de Normalisation. 2003d. Bituminous mixtures – Test methods for hot mix asphalt – Part 23: Determination of the indirect tensile strength of bituminous specimens. *EN 12697-23:2003*. London: BSI; Berlin: DIN; Paris: AFNOR; and other European standards institutions.

Comité Européen de Normalisation. 2004. Bituminous mixtures – Test methods for hot mix asphalt – Part 17: Particle loss of porous asphalt specimen. *EN 12697-17:2004*. London: BSI; Berlin: DIN; Paris: AFNOR; and other European standards institutions.

Comité Européen de Normalisation. 2010b. Bituminous mixtures – Test methods for hot mix asphalt – Part 44: Crack propagation by semi-circular bending test. *EN 12697-44:2010*. London: BSI; Berlin: DIN; Paris: AFNOR; and other European standards institutions.

Comité Européen de Normalisation. 2012h. Bituminous mixtures – Test methods for hot mix asphalt – Part 34: Marshall test. *EN 12697-34:2012*. London: BSI; Berlin: DIN; Paris: AFNOR; and other European standards institutions.

Comité Européen de Normalisation (CEN). 2012i. Bituminous mixtures – Test methods for hot mix asphalt – Part 20: Indentation using cube or cylindrical specimens (CY). *BS EN 12697-20:2012*. London: BSI; Berlin: DIN; Paris: AFNOR; and other European standards institutions.

Comité Européen de Normalisation (CEN). 2012j. Bituminous mixtures – Test methods for hot mix asphalt – Part 21: Indentation using plate specimens. *BS EN 12697-21:2012*. London: BSI; Berlin: DIN; Paris: AFNOR; and other European standards institutions.

Comité Européen de Normalisation. 2012k. Bituminous mixtures – Test methods for hot mix asphalt – Part 46: Low temperature cracking and properties by uniaxial tension tests. *EN 12697-46:2012*. London: BSI; Berlin: DIN; Paris: AFNOR; and other European standards institutions.

Comité Européen de Normalisation. 2016a. Bituminous mixtures – Material specifications – Part 1: Asphalt concrete. *EN 13108-1:2016*. London: BSI; Berlin: DIN; Paris: AFNOR; and other European standards institutions.

Comité Européen de Normalisation. 2016b. Bituminous mixtures – Material specifications – Part 7: Porous asphalt. *EN 13108-7:2016*. London: BSI; Berlin: DIN; Paris: AFNOR; and other European standards institutions.

Comité Européen de Normalisation. 2016f. Bituminous mixtures – Material specifications – Part 6: Mastic asphalt. *EN 13108-6:2016*. London: BSI; Berlin: DIN; Paris: AFNOR; and other European standards institutions.

Comité Européen de Normalisation. 2016j. Bituminous mixtures – Test methods for hot mix asphalt – Part 25: Cyclic compression test. *EN 12697-25:2016*. London: BSI; Berlin: DIN; Paris: AFNOR; and other European standards institutions.

Comité Européen de Normalisation. 2016k. Bituminous mixtures – Test methods for hot mix asphalt – Part 16: Abrasion by studded tyres. *EN 12697-16:2016*. London: BSI; Berlin: DIN; Paris: AFNOR; and other European standards institutions.

Hunter, R N, A Self and J Read. 2015. *The Shell Bitumen Handbook*. 6th edition. London: ICE Publishing.

Kandhal, P S and L A Cooley. 2001. The restricted zone in the Superpave aggregate gradation specification. *NCHRP Report 464*. Washington, DC: National Academy Press. http://onlinepubs.trb.org/Onlinepubs/nchrp/nchrp_rpt_464-a.pdf

Minnesota Department of Transportation. 2016. Low temperature cracking in asphalt pavements. www.dot.state.mn.us/mnroad/projects/Low_Temp_Cracking/

Nicholls, J C, C Roberts and P Samuel. 2006a. Implications of implementing the European asphalt test methods. *TRL Report TRL656*. Wokingham: TRL Limited.

Nikolaides, A. 2015. *Highway Engineering – Pavements, Materials and Control of Quality*. London: CRC Press.

Szatkowski, W S and F A Jacobs. 1977. Dense wearing courses in Britain with high resistance to deformation. In *Colloquium 77, Plastic Deformability of Bituminous Mixes*, pp. 65–67. Zurich.

Thom, N. 2014. *Principles of Pavement Engineering*. 2nd edition. London: ICE Publishing.

Chapter 6

Structural properties

6.1 STIFFNESS

6.1.1 Issue being addressed

Stiffness is used as the principal parameter for designing pavements with the ability of the pavement to spread the load being proportional to both the stiffness and the depth, so that the thickness can be reduced as the stiffness is increased. This relationship led to a 'great Pascal race', in the United Kingdom, in which everybody was trying to save money by maximising the stiffness. However, this approach to pavement design risks premature failure if the stiffness is not up to the design value when constructed or if there is any loss of strength for any reason. In general, the stiffness will be expected to increase with time from binder hardening, but the stiffness can be significantly and adversely affected by high moisture susceptibility or other weathering effects.

6.1.2 Conventional measures

In the United Kingdom, the stiffness of the mixture is defined in HD 26/06 (Highways Agency et al., 2006b) by the mixture type for use in nomographs, where:

> A DBM125 base is the least stiff material, and so requires the thickest construction. The stiffness of asphalt material then increases from HRA50, through DBM50/HDM50, to EME2.

Effectively, the mixture type and bitumen grade (all the mixtures listed include a binder grade, with the harder/lower penetration grade bitumen assumed to give greater stiffness) are used to provide conventional requirements that avoid the temptation of overestimating the stiffness that would routinely be produced.

6.1.3 Stiffness modulus (fundamental)

There are several different ways of stressing specimens to provide an assessment of the stiffness modulus of an asphalt mixture. The different methods do not provide identical results, but they are generally close enough for them to all be considered fundamental tests, although those differences do indicate they cannot be truly fundamental. The European test standard, EN 12697-26 (CEN, 2012l), includes six methods for characterising the stiffness modulus of asphalt mixtures, including bending tests and direct and indirect tensile tests. These methods are described in Table 6.1. The stiffness modulus will be dependent on both the test temperature and load frequency.

The results from these different methods on samples of the same mixture will not be identical, but they are sufficiently similar for EN 13108 to regard them as equivalent.

Pavements with asphalt that have greater values of stiffness modulus allow for either thinner layer thickness or improved certainty of performance. The categories of stiffness listed in EN 13108 for use in Europe are for minimum stiffness moduli of between 1,500 MPa and 21,000 MPa for asphalt concrete and stone mastic asphalt and between 500 MPa and 21,000 MPa for hot rolled asphalt plus no requirement categories. In order to control the performance of new asphalt laid against existing construction, categories for maximum stiffness moduli are also given of between 7,000 MPa and 30,000 MPa for asphalt concrete and stone mastic asphalt

Table 6.1 CEN methods for determining stiffness modulus

Name	Symbol	Description
Two-point bending test on trapezoidal specimens	2PB-TR	A specimen is fixed at one end loaded at the other end as a cantilever by a sinusoidal force or a sinusoidal deflection
Two-point bending test on prismatic specimens	2PB-PR	
Three-point bending test on prismatic specimens	3PB-PR	A prismatic specimen is subjected to three-point or to four-point sinusoidal bending with free rotation at all load and reaction points
Four-point bending test on prismatic specimens	4PB-PR	
Indirect tension test on cylindrical specimens	IT-CY	An indirect tensile load is applied to an asphalt cylinder to determine the elastic stiffness modulus
Direct tension–compression test on cylindrical specimens	DTC-CY	A sinusoidal strain is applied on a cylindrical sample glued on two steel plates screwed to a loading rig
Direct tension test on cylindrical specimens	DT-CY	Uniaxial tensile loads are applied to a specimen at given temperatures and loading times
Direct tension test on prismatic specimens	DT-PR	
Cyclic indirect tension test on cylindrical specimens	CIT-CY	Cyclic indirect tensile loads are applied to an asphalt specimen

and between 1,500 MPa and 30,000 MPa for hot rolled asphalt (plus no requirement categories).

The precision given in EN 12697-26 are a repeatability of 335 MPa and a reproducibility of 2,740 MPa for the two-point bending test on trapezoidal specimens. No precision data are given for the other options.

6.1.4 Specifying stiffness

Stiffness is a property required for pavement design, so there is no need to specify the property for asphalt to be used for non-designed pavements. For designed pavements, the conventional measures are all that is needed if the pavement design method is based around those measures while the actual stiffness will need to be determined for more sophisticated pavement design methods, whether to determine that input parameter or to confirm that adequate stiffness has been provided. The choice of method to determine the stiffness value is generally not critical, but it is probably preferable to choose the test method used to develop the pavement design model.

6.2 FATIGUE RESISTANCE

6.2.1 Issue being addressed

Fatigue occurs when the asphalt cannot resist the cumulative effect of repeated loads and generally results in the cracking and even disintegration of the layer being loaded and any higher layers. Fatigue resistance is generally achieved by the material being more flexible so that it can sustain repeated deformations, but stiffer materials will deform less under the same loads and, therefore, be damaged to a lesser extent. Unfortunately, the two aspects are effectively mutually exclusive with resistance to fatigue generally being considered as the ability to absorb repeated deformations rather than minimising size.

The resistance to fatigue, alongside stiffness modulus, has long been the primary consideration in mixture design. In the French method for design of asphalt mixtures (Delorme et al., 2007), there are four levels with fatigue at the top, as shown in Figure 6.1 and Table 6.2 (Widyatmoko et al., 2007).

The levels used for a mixture design depend on the loading, environmental conditions and the performance requirements needed for the pavement design. The tests in Level 1 are mandatory for all mixture designs; Level 2 requirements are additional for layers in pavements that will be subjected to high traffic; Level 3 requirements are for base and binder course mixtures when the determination of the stiffness modulus of the mixture is required for pavement design purposes and Level 4 requirements are for all heavily trafficked pavements for mixtures used in base layers of new pavements or of overlays.

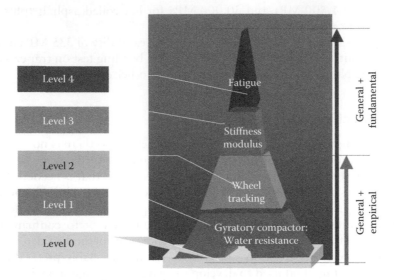

Figure 6.1 French pyramid of design levels. (Adapted from Delorme, J-L, C de la Roche and L Wendling. 2007. *Manuel LPC d'aide à la formulation des enrobés [LPC Bituminous Mixtures Design Guide].* Paris: Laboratoire Central des Ponts et Chaussées.)

Table 6.2 Four levels of testing

Description of tests performance	Requirement	Level 1	Level 2	Level 3	Level 4
Gyratory shear compactor	Workability				
Duriez at 18°C	Durability				
Rutting test	Deformation				
Mechanical characterisation tests with stiffness/complex modulus or direct tensile	Load-bearing capacity				
Fatigue test	Fatigue cracking				

Source: Widyatmoko et al. 2007. The use of French asphalt materials in UK airfield pavements. In *23rd PIARC World Road Congress Paris,* 17–21 September 2007. Paris: PIARC – World Road Association. www.researchgate.net/publication/288824215_The_Use_of_French_Airfield_Asphalt_Concrete_in_the_UK

However, the resistance to fatigue has been found to be necessary for thick pavements when designed as long-life pavements (Nunn et al., 1997) which are called, more optimistically, perpetual pavements in America. It has been found that most thick pavements either maintain their strength or become stronger over time, rather than gradually weakening with trafficking as previously assumed, and that any deterioration that occurs will generally occur in the surfacing rather than the structural layers. Furthermore, cracking, including fatigue and reflective, have been found to originate at the surface and travel downward, even if the cause of the crack is lower down, rather than start at the bottom and come up. Therefore, the surfacing has to be replaced before any cracks penetrate the structural layers for the pavement to be revitalised with no need for full-depth replacement.

Therefore, with long-life pavements, theoretically there is no need to consider resistance to fatigue. However, some counter is needed to stiffness modulus otherwise the mixture design will produce an over-harsh mixture that will not have sufficient durability (Chapter 7).

Fatigue resistance can be undertaken in either controlled stress or controlled strain modes. Controlled stress is where the load remains constant with the resulting deformation increasing as the mixture weakens with fatigue while controlled strain is where the deformation produced remains constant so that the load applied has to be reduced as the mixture weakens with fatigue. It is generally assumed that thicker pavements will follow the controlled strain mode because any deformation will be restricted, whilst thinner pavements will follow the controlled stress mode. In practice, a pavement will be fatigued in a mixture of both modes.

As well as the two modes, the loads that can be repeated are numerous, including bending, shear and torsional forces. Unfortunately, the resistance to fatigue in the different modes against the various types of load differs markedly, making it difficult to define a single criterion. In particular, the ranking of different mixtures tends to be reversed when changing from controlled stress to controlled strain, or vice versa.

6.2.2 Conventional measures

The simplest measure to obtain good resistance to fatigue is to require a high binder content. A slightly more sophisticated approach is to require a minimum binder film thickness, as is done through the mechanism of requiring a minimum binder richness modulus in the mixture design of EME mixtures when fatigue testing is omitted. The binder richness modulus is calculated as

$$K = \frac{B_{PPC} \times \rho_{mc}}{2650 \times \sqrt[5]{(0.25G + 2.3S + 12s + 135f)}}$$

where

K = binder richness modulus

B_{PPC} = mass of soluble binder as a proportion of total dry mass of aggregate, including filler (%)

ρ_{mc} = theoretical aggregate density (in Mg/m³)

G = proportion by mass of aggregate particles over 6.3 mm (decimal fraction)

S = proportion by mass of aggregate particles between 6.3 and 0.315 mm (decimal fraction)

s = proportion by mass of aggregate particles between 0.315 and 0.08 mm (decimal fraction)

f = proportion by mass of aggregate particles smaller than 0.08 mm (decimal fraction)

6.2.3 Resistance to fatigue (fundamental)

As with stiffness modulus, there are several different ways of fatiguing specimens to provide an assessment of the fatigue resistance of an asphalt mixture. The different methods do not provide identical results or even rank mixtures in the same order, but they are still often considered as fundamental tests. The European test standard, EN 12697-24 (CEN, 2012m), includes five methods for characterising the resistance to fatigue of asphalt mixtures, including bending tests and direct and indirect tensile tests. These methods are described in Table 6.3.

With all the tests, the criterion for defining when the sample has failed due to fatigue is required. The conventional criterion of failure in fatigue is the number of load applications required for the stiffness modulus of the asphalt to be reduced to half its initial value, although other criteria can be used.

Because of the different results, only two of the options are accepted for CE Marking of asphalt products to EN 13108-20 (CEN, 2016l), the two-point bending test on trapezoidal specimens and four-point bending test on prismatic specimens, with the selection being nominally dependent on the mixture design procedure being used.

Lower strains after 10^6 cycles and reduced slopes of the fatigue line indicate a greater resistance to fatigue. EN 13108 has categories for the

Table 6.3 CEN methods for determining fatigue resistance

Name	Description
Two-point bending test on trapezoidal specimens	Repeated loading with controlled strain is applied to a specimen that is held at one end and loaded at the other to produced cantilever bending
Two-point bending test on prismatic specimens	Repeated bending loading with controlled strain is applied to a specimen that is held at one end and loaded at the other to produce cantilever bending
Three-point bending test on prismatic specimens	Repeated bending loading with controlled strain is applied to a specimen
Four-point bending test on prismatic specimens	Repeated bending loading with controlled strain is applied to a specimen
Indirect tension test on cylindrical-shaped specimens	Repeated loading with controlled stress is applied to a specimen

strain after 10^6 cycles with minima of from 50 to 310 µstrain for asphalt concrete and stone mastic asphalt. In addition for asphalt concrete, there are categories for the number of load cycles to macro-crack formation with minima of from 3,000 to 45,000 and from 30,000 to 800,000 for quality indices of 0.1‰ and 0.05‰, respectively, plus no requirement categories.

The precision given in EN 12697-24 for the two-point bending test on trapezoidal specimens are

- A repeatability of 4.2 µstrain and a reproducibility of 8.3 µstrain for the strain after 10^6 cycles
- A repeatability of 0.06062 and a reproducibility of 0.0642 for the slope of the fatigue line

Precision data are not given for the other options other than a confidence interval of the strain for the indirect tensile test on cylindrical-shaped specimen option.

6.2.4 Specifying against fatigue

With the concept of long-life pavements, fatigue resistance could be regarded as superfluous because it is unnecessary for long-life pavements and too expensive for minor pavements. However, there are pavements that are not long-life but are designed such that fatigue resistance should be a consideration and, in any case, there does need to be some balance against designing mixtures just for mechanical strength and so making them susceptible to serviceability problems.

For most pavements, conventional measures should provide the balance. When the measurement of fatigue resistance is required as part of a pavement design method, the test method should be the one against the design method was calibrated; the significant differences in the methods mean that, unlike with stiffness, an alternative method may not be applicable.

REFERENCES

Comité Européen de Normalisation. 2012l. Bituminous mixtures – Test methods for hot mix asphalt – Part 26: Stiffness. *EN 12697-26:2012*. London: BSI; Berlin: DIN; Paris: AFNOR; and other European standards institutions.

Comité Européen de Normalisation. 2012m. Bituminous mixtures – Test methods for hot mix asphalt – Part 24: Resistance to fatigue. *EN 12697-24:2012*. London: BSI; Berlin: DIN; Paris: AFNOR; and other European standards institutions.

Comité Européen de Normalisation. 2013a. Bituminous mixtures – Test methods for hot mix asphalt – Part 3: Bitumen recovery – Rotary evaporator. *EN 12697-3:2013*. London: BSI; Berlin: DIN; Paris: AFNOR; and other European standards institutions.

Comité Européen de Normalisation. 2016l. Bituminous mixtures – Material specifications – Part 20: Type testing. *EN 13108-20:2016*. London: BSI; Berlin: DIN; Paris: AFNOR; and other European standards institutions.

Delorme, J-L, C de la Roche and L Wendling. 2007. *Manuel LPC d'aide à la formulation des enrobés [LPC Bituminous Mixtures Design Guide]*. Paris: Laboratoire Central des Ponts et Chaussées.

Nunn, M E, A Brown, D Weston and J C Nicholls. 1997. Design of long-life flexible pavements for heavy traffic. *TRL Report 250*. Wokingham: TRL Limited.

Spiegl, M. 2007. Tieftemperaturverhalten von bitumiösen baustoffen – Labortechnische ansprache und numerische simulation des gebrauchsverhaltens (Low temperature behaviour of bituminous construction materials – Assessment on laboratory scale and numerical simulation of the low temperature performance). Doctoral thesis. Techische Universtität Wien.

The Highways Agency, Transport Scotland, Welsh Assembly Government and The Department for Regional Development, Northern Ireland. 2006b. Pavement design. In *Design Manual for Roads and Bridges: Volume 7, Pavement Design and Maintenance: Section 2, Pavement Design and Construction: Part 3, HD 26/06*. London: The Stationery Office. www.standardsforhighways.co.uk/dmrb/vol7/section5/hd3606.pdf

Widyatmoko, I, B Hakim, C Fergusson and J Richardson. 2007. The use of French asphalt materials in UK airfield pavements. In *23rd PIARC World Road Congress Paris*, 17–21 September 2007. Paris: PIARC – World Road Association. www.researchgate.net/publication/288824215_The_Use_of_French_Airfield_Asphalt_Concrete_in_the_UK

Chapter 7

Serviceability properties

7.1 SERVICEABILITY

The lack of serviceability due to the effects of ageing, water, chemicals or other outside influences lead to a diminution of other structural properties that will result in the actual failure of the pavement.

7.2 MOISTURE DAMAGE

7.2.1 Issue being addressed

Most aggregates are classified as hydrophilic, or water loving, and oleophobic, or oil hating (Hunter et al., 2015). As such, bitumen has a tendency to be stripped from the aggregate particles when there is water present that can replace the bitumen on the particle surface. The extent to which this phenomenon will occur is also dependent on the aggregate type, with acidic aggregates, examples being quartz and granite, being more difficult to coat with bitumen and more likely to strip than basic aggregates, such as basalt and limestone (Hunter et al., 2015).

The propensity to strip will also be dependent on the regularity that water is present, the time period that the water remains there and the extent to which the binder film on the particle is complete to stop the water ever reaching the aggregate surface. Therefore, the extent to which the moisture sensitivity needs to be addressed will depend on the site where the asphalt is to be laid, with joints often being the locations where water can remain longest after rainfall.

Any stripping will not only lead to loss of aggregate particles but also a reduction in several other asphalt properties, particularly those related to strength. The reduction in strength can result without visible signs of aggregate particles no longer being fully coated. In extreme cases, the presence of water can completely destroy the cohesion of an asphalt mixture (Figure 2.3).

The typical methods for assessing the potential for moisture damage are to test either

- The affinity between the component materials of the asphalt
- The change in a physical property of the asphalt mixture following a period of immersion

Often both types of test are specified when there are particular concerns about moisture damage.

7.2.2 Conventional measures

The basic conventional measures to avoid moisture damage are a minimum binder film thickness, or binder content as its surrogate, and the choice of aggregate type. The addition of adhesion agents can assist when the aggregate type is not ideal, and sometimes is specified as a requirement irrespective of aggregate type if the situation where the asphalt is to be laid is particularly prone to moisture damage.

7.2.3 Aggregate/binder affinity (simulative)

7.2.3.1 Static method

A number of single-sized aggregate particles that have been coated in bitumen are left in a tray whilst immersed under distilled water for a specific time period. The particles are then visually assessed for the proportion of particles that are no longer completely coated. The parameters that need to be set for this simulative test are aggregate size, quantity of binder used for coating, minimum number of observations, temperature of the water and time period of the immersion. The test is simple and capable of dealing with highly abrasive aggregates but the assessment is subjective, so precision is limited. The static immersion method for aggregate/binder affinity is standardised in Europe as Clause 6 of EN 12697-11 (CEN, 2012n) and in America as AASHTO T-182 (AASHTO, 1984).

Fewer incompletely covered aggregate particles indicate improved aggregate/binder affinity. However, there are no categories in EN 13108 for specifying aggregate/binder affinity. EN 12697-11 gives no precision data for this test.

7.2.3.2 Rolling bottle method

The rolling bottle method is similar to the static immersion method except that the coated particles are rolled in a bottle whilst in the presence of water instead of being left static whilst immersed under water. Also, the visual assessment is extended to estimate the proportions that are still coated on each particle. The test is again relatively simple but subjective and, hence, not precise. However, the number of single-sized aggregate particles that have been coated in bitumen and rolled makes the test unsuitable for highly abrasive aggregates because the particles will abrade each other. The rolling bottle method for aggregate/binder affinity is standardised in Europe as Clause 5 of EN 12697-11 (CEN, 2012n).

Fewer incompletely covered aggregate particles indicate improved aggregate/binder affinity. However, there are no categories in EN 13108 for specifying aggregate/binder affinity. EN 12697-11 gives the precision for the rolling bottle test of a repeatability of 20% and a reproducibility of 30%.

7.2.3.3 Boiling water stripping method

In the boiling water stripping method, water in a beaker is brought to boiling point and then single-sized aggregate particles that have been coated in bitumen are added and any bitumen that floats to the surface is removed before the water is brought back to the boil for a limited period. Any further bitumen floating to the surface is removed and the aggregate sieved out and allowed to cool. The particles are put into contact with hydrochloric acid (calcareous aggregates) or hydrofluoric acid (silico-calcareous or siliceous aggregates) for a time and then the acid removed from the aggregate. The volume of acid consumed in the process is determined by repeated titrations with sodium hydroxide (calcareous aggregates) or potassium hydroxide (calcareous aggregates) in the presence of phenolphthalein. The test is objective with high precision and can be used for any binder/aggregate combination in which the mineral aggregate is calcareous, silico-calcareous or siliceous. However, the use of dangerous acids raises significant health and safety issues that make it impractical as a routine test. The boiling water method for aggregate/binder affinity is standardised in Europe as Clause 7 of EN 12697-11 (CEN, 2012n) and in America as ASTM D3625 (ASTM, 2012b).

Fewer incompletely covered aggregate particles indicate improved aggregate/binder affinity. However, there are no categories in EN 13108 for specifying aggregate/binder affinity. For precision, EN 12697-11 only gives a repeatability coefficient of variation of 15% for a level of 2% and no reproducibility.

7.2.4 Water sensitivity (simulative)

The water sensitivity involves comparing the results of a physical test on samples with and without having gone through a conditioning phase. The conditioning usually consists of being immersed in water for a specific time at a specific temperature. If the physical test is destructive, different samples are needed for the conditioned and the unconditioned measurements, which add potential variability to the results. The European test, EN 12697-12 (CEN, 2008c), has methods using alternative two destructive physical tests, indirect tensile strength (ITS) and compression strength, while the UK Highway Authorities Products Approval Scheme (HAPAS) for thin surface course systems used the non-destructive indirect tension stiffness test on cylindrical specimens. EN 12697-12 also includes a separate method for soft asphalt mixtures using the bond between the bitumen and aggregate.

Higher proportions of the property retained demonstrate that water has less deleterious effect on that property. EN 13108 gives categories of either indirect tensile strength ratio (ITSR) or compressive strength ratio for use in Europe with minima of

- 60% to 90% in 5% increments plus a no requirement category for asphalt concrete and SMA
- 65% to 90% in 5% increments plus a no requirement category for BBTM
- 60% to 90% in 10% increments plus a no requirement category for soft asphalt
- 60% to 80% in 5% increments plus a no requirement category for hot rolled asphalt
- 50% to 95% in 5% increments plus a no requirement category for porous asphalt
- 75% to 95% in 5% increments plus a no requirement category for AUTL

No precision data are given in EN 1267-12.

7.2.5 Saturation ageing tensile stiffness conditioning (simulative)

The saturation ageing tensile stiffness (SATS) was developed to assess the durability of adhesion in base and binder course asphalt mixtures following some major failures on site. The failed mixtures, and therefore the test, were specific to asphalt concrete with binder contents between 3.5% and 5.5% of 10/20 pen hard paving grade bitumen and air voids contents between 6% and 10%. The European standard for this test is EN 12697-45 (CEN, 2012o) although alternative conditions for mixtures with binders other than 10/20 hard grade bitumen or other situations are being developed (Grenfell et al., 2011; Nicholls et al., 2011).

Cylindrical specimens are subjected to moisture saturation by a vacuum system before being transferred into a pressurised vessel partially filled with water where they are subjected to a conditioning procedure by storage at elevated temperature and pressure for a specific time. The average ratio of the indirect tensile stiffness before and after conditioning by storage under increased pressure and elevated temperature on the specimens above the water defines the sensitivity of the material to ageing and moisture.

Higher proportions of the retained indirect tensile stiffness demonstrate that the ageing simulation has less deleterious effect on the stiffness. EN 13108-1 (CEN, 2016a) gives categories of minima retained stiffness of 60% to 100% in 10% increments for AC plus no requirement categories. EN 12697-45 gives no precision data.

7.2.6 Moisture-induced stress test (simulative)

Moisture conditioning protocols often both fail to capture the time frame over which moisture infiltration occurs and disregard the short-term moisture processes related to pumping action (Kringos et al., 2009). In addition, field observations have shown that stripping of open asphalt mixtures is a localised phenomenon in trafficked areas of a pavement which are oversaturated with water (Kandhal et al., 1989). Pumping action can be an important damage mechanism which acts concurrently with the long-term damage processes and contributes to premature failure in asphalt pavements.

The moisture-induced stress test (MIST) is a moisture conditioning protocol that is designed to distinguish the individual contributions of short- and long-term moisture damage to mixture degradation (Varveri et al., 2014). The evaluation of the asphalt mixtures for their sensitivity to moisture is performed on the basis of their ITS and ITSR.

The MIST equipment (Figure 7.1) is a self-contained unit that includes a hydraulic pump and a piston mechanism that is designed to cyclically apply pressure inside a sample chamber. Moisture conditioning is performed by placing a compacted asphalt sample in the chamber and filling it with water. Then, the water is pushed and pulled through the sample, creating pressure cycles between zero and a specific pressure. The parameters that can be chosen are pressure, temperature and number of conditioning cycles to replicate different combinations of traffic and environmental conditions.

7.2.7 Specifying against moisture damage

The conventional measures are adequate for most situations. However, specification to minimise the potential for moisture damage is important when the location is more exposed than usual, particularly if the pavement is liable to regular flooding or immersion, or if the aggregate sources generally used in the area are known to have limited affinity to bitumen. The European specification then uses the water sensitivity, leaving the aggregate/binder affinity test for use by the supplier despite the fact the latter test determines the property (binder 'stickability' to the aggregate particles) while the former looks at the result of the property (loss of performance for other properties).

When the specification is to limit the potential for moisture damage, it is the water sensitivity test that is usually called upon, typically with the

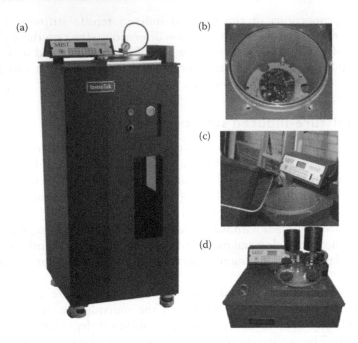

Figure 7.1 Moisture-induced sensitivity tester. (a) MIST machine; (b) chamber with specimen; (c) chamber sealed from top; (d) CoreDry equipment. (Adapted from Varveri, A. et al. 2014. Laboratory study on moisture and ageing susceptibility characteristics of RA and WMA mixtures. *EARN deliverable D7.* www.trl.co.uk/solutions/road-rail-infrastructure/sustainable-infrastructure/earn/)

tensile strength ratio rather than the compressive strength ratio. The rolling bottle and static options of the aggregate/binder sensitivity test are suitable for specifying moisture damage, with the static option being routinely used for airfield pavements. The third aggregate/binder sensitivity test, the boiling water stripping option, is not suitable for routine specification because they involve the use of hazardous acids. It is, however, suitable as a reference method. The SATS test is designed for a specific material type and is not appropriate for other situations while the MIST test has great potential but has not been fully standardised.

7.3 RESISTANCE TO CHEMICALS

7.3.1 Issue being addressed

Whilst moisture can cause damage to asphalt, petroleum-based hydrocarbons and aggressive chemicals can cause even more damage. The hydrocarbons, which include the various commonly used engine fuels and lubricants, are damaging because they will mix with the bitumen, being from the same source of crude oil, and leave it softer and, hence, make the asphalt mixture less stiff and less deformation resistant. The aggressive chemicals include acids and deicing fluids, the latter being more commonly found around airfield and highway pavements.

The need to ensure that asphalt is resistant to hydrocarbons is greatest in areas where vehicles are refuelled or maintained. The potential damage is the reason for the forecourts of most petrol stations being paved with concrete or block pavers. However, the area where aircraft are refuelled on an airfield is less restricted and the asphalt needs to be resistant to fuel when the pavement is of flexible construction. Similarly, in snowy conditions, deicing fluids rather than salt or grit tend to be used more frequently on airfield than highway pavements. Therefore, requirements for resistance to fuel and/or deicing fluids tend to be more common for airfields than highways.

Logically, similar tests could be used for resistance to fuel and/or deicing fluids as for resistance to water but with the water being replaced by the aggressive fluids, although there may need to be extra precautions for some fluids. However, the tests for these fluids are often different from those for moisture damage and even from each other. In the case of the European test methods, the tests for fuel and deicing fluids were dictated by the CEN airfield group and are totally different, making it unnecessarily awkward and costly for laboratories to be set up for both tests.

7.3.2 Resistance to fuel (simulative)

The European Standard for the resistance of an asphalt mixture to fuels is EN 12697-43 (CEN, 2005). A test specimen, either laboratory-prepared or cored from a pavement, is initially soaked in the relevant fuel for a specific time, which is longer for mixtures with polymer-modified bitumen than for those with paving grade bitumen, before being brushed with a steel brush mounted in a mixer for another specific time. The material lost from the sample during each phase is used to assess the resistance to that fuel for that asphalt mixture. The test is normally carried out with jet fuel for airfield use.

Lower proportions of the material lost demonstrate that the fuel has less deleterious effect on the asphalt. EN 13108 gives categories for resistance to fuel for use on European airfields with maxima loss of 6% to 15% in 1% increments for AC, BBTM and SMA and of 1% to 8% in 1% increments for HRA, MA, PA and AULT plus a no requirement category for all. EN 12697-43 gives no precision data for the test.

7.3.3 Resistance to deicing fluids (simulative)

The resistance to deicing fluids is an issue primarily for airfield pavements because salt or brine are used on highways rather than deicing fluids.

The European Standard for the resistance of an asphalt mixture to deicing fluids, including solutions of acetate and formate, is EN 12697-41 (CEN, 2013b). Cylindrical specimens have well-defined test surfaces drilled out before half of them are stored in the relevant deicing fluid for a specific period and the other half left unconditioned. A steel plate is then bonded to the test surface of each specimen in turn pulled off with a tensile force applied perpendicular at a constant rate. The mean ratio of conditioned and unconditioned tensile forces at failure load is used to assess the resistance to that deicing fluid for that asphalt mixture.

Higher proportions of tensile force that is retained demonstrate that the deicing fluid has less deleterious effect on the asphalt. EN 13108 gives categories with minima retained strength of 55% to 100% in 15% increments plus a no requirement category for all asphalt mixture types apart from soft asphalt. EN 12697-41 gives precision data for the test of a repeatability 360 N and a reproducibility of 620 N.

7.3.4 Specifying for resistance to chemicals

Most situations do not require asphalt to be specifically specified as resistant to chemicals. The locations where resistance to fuel is important is where vehicles are regularly refuelled, such as petrol station forecourts, or where fuel is likely to be spilled, including parking areas for aircraft. Resistance to deicing fluids is needed where the pavement is regularly deiced with such fluids each winter. However, materials other than asphalt that are inherently more resistant are widely used for the surface course in those locations, such as concrete for petrol station forecourts.

7.4 IMPERMEABILITY

7.4.1 Issue being addressed

Traditionally, the requirement was for impermeability to be maximised as much as practicable in order to keep moisture out of the pavement because of the damage that it can do to asphalt (Section 7.2). One problem with this approach is that, if the material is impermeable, any moisture that does get into the pavement, whether through the limited permeability or through joints and/or damage, is effectively trapped there. In more recent times, the concept of sustainable urban drainage systems (SUDS) requires the asphalt to be sufficiently permeable for any water to pass through the pavement. Whilst SUDS is not yet practical for major highways or airfield pavements, it has been used on parking areas and minor roads to allow development without significantly upsetting the water table or requiring large quantities of surface water to have to be drained during wet weather conditions. There is also an intermediate situation where the surface course only is permeable, but then a drainage system needs to be in place to take the water away from the interface between the permeable surface course and the impermeable binder course (Nicholls and Carswell, 2001). Nevertheless, asphalt is either required to be as impermeable as possible or as permeable as possible!

One problem is that both the impermeability and permeability will change more dramatically with time than other properties. Any damage to the asphalt, particularly cracking or fretting, will reduce the permeability while open-textured mixtures, in particular porous asphalt, will clog with detritus very quickly such that the initial permeability cannot be fully recovered even with cleaning (Nicholls, 1997).

7.4.2 Conventional measures

The simplest means of increasing the permeability or impermeability of asphalt is by the choice of the asphalt type, with the range of values of permeability that can be expected from some common mixture types being given in Table 7.1 (Daines, 1995).

In general, permeability can be reduced by increasing the binder and/or filler contents to effectively fill up the voids, but such measures are rarely specified for this purpose. Also, the compaction needs to be thorough in order to achieve good impermeability, but permeability gained through limiting the compaction will be short lived because of the adverse effect on the durability of under-compacted mixtures.

7.4.3 Air voids content (surrogate)

The usual test for impermeability, as opposed to permeability, is the air voids content. However, the air voids content is only a surrogate for permeability because the extent that the voids are interconnecting as well as their content will affect the permeability. Furthermore, it is the air voids content that is measured whereas it is the impermeability to water that is the

Table 7.1 Relative permeability of surfacing materials

Material	Typical air voids content (%)	Approximate water permeability (m/s)
Mastic asphalt	<1	$<10^{-11}$
Hot-rolled asphalt (30% stone)	2–8	$10^{-11}–10^{-10}$
Hot-rolled asphalt (55% stone)	2–6	$10^{-11}–10^{-10}$
Asphalt concrete	3–5	$10^{-10}–10^{-8}$
Close-graded bitumen macadam	4–7	$10^{-8}–10^{-5}$
Open-graded bitumen macadam	12–20	$10^{-8}–10^{-3}$
Porous asphalt	15–25	$10^{-4}–10^{-2}$

Source: Adapted from Daines, M E. 1995. Tests for voids and compaction in rolled asphalt surfacings. TRL Project Report PR78. Wokingham: TRL Limited.

fundamental property. The size of the voids and interconnecting passage will affect the fluids that can pass through them, with the viscosity and surface tension effects making some air voids impassable to water.

The measurement of air voids content is covered in Section 3.4.1.

7.4.4 Permeability (simulative)

The permeability of an open asphalt mixture can be measured in the laboratory on either laboratory-prepared samples or cores taken from site. A column of water with a constant height is applied to a cylindrical specimen and is allowed to permeate through the specimen for a controlled time. The direction of flow being measured can be either vertical or horizontal. The European standard for permeability is EN 12697-19 (CEN, 2012p).

The greater the permeability, the easier it is for water to flow through the mat when that is required for a permeable pavement. EN 13108-7 (CEN, 2016b) gives categories for specifying for both horizontal and horizontal permeabilities with minima of $0.1 \cdot 10^{-3}$ and $0.5 \cdot 10^{-3}$ m/s to $4.0 \cdot 10^{-3}$ m/s in $0.5 \cdot 10^{-3}$ m/s increments plus no requirement categories. EN 12697-19 gives no precision data for the test.

7.4.5 Hydraulic conductivity (simulative)

The *in situ* drainability of an open asphalt mixture can be evaluated as the time that a fixed volume of water takes to dissipate through a permeameter into an annular area of the surface course of a pavement under known head conditions. The relative hydraulic conductivity is the reciprocal of the outflow time; the result is relative, rather than absolute, because the time taken is dependent on the dimensions of the permeameter. The test measures the ability to drain water (drainability) of a surface course and can be used as a compliance check to ensure that a permeable surface course has the required properties when it is laid and subsequently to establish the change of drainage ability with time. The relative hydraulic conductivity is not identical to permeability because the water can emerge from the pavement immediately around the permeameter or further away, depending on a number of factors. The European test for drainability is EN 12697-40 (CEN, 2012g).

Because hydraulic conductivity is an *in situ* test and the EN 13108 series is only intended for asphalt 'in the back of the lorry', EN 13108-7 (CEN, 2016b) has no categories for specifying hydraulic conductivity. EN 12697-40 gives no precision data for the test.

7.4.6 Specifying impermeability or permeability

The requirement of impermeability is usually achieved using conventional measures and confirmed, when necessary, by specifying the air voids content. For most situations, explicit specification of impermeability is not necessary, leaving such specification to asphalt for critical situations. When specifying permeability, which does need to be explicitly specified when required, the horizontal and/or vertical permeability can be used for the potential permeability in the laboratory while the hydraulic conductivity/drainability can be used to confirm that it has been achieved *in situ*.

7.5 AGEING POTENTIAL

7.5.1 Issue being addressed

Several of the properties related to durability that are used involve testing the mixture with the properties that it has when constructed. However, the binder in asphalt does change its properties with time and exposure to oxygen, ultraviolet light, chemicals and various other influences. Therefore, the properties of the asphalt will also change with, say, the resistance to deformation improving and fatigue resistance deteriorating as the asphalt hardens. Ideally, this change should be included in the assessment of mixtures, as is required in Mandate M 124 to CEN/CENELEC concerning the execution of standardisation work for harmonised standards on road construction products.

The mandate includes the need to address durability 'against ageing, weathering, oxidation, wear, chemicals, wear of studded tyres, stripping, ... as relevant' for all the properties of asphalt for road construction. Unfortunately, ageing procedures that age the asphalt appropriately have been developed but have not been widely accepted as representative. Furthermore, there also needs to be a methodology to assess the expected damage from each property from its initial and aged states.

7.5.2 Binder ageing (surrogate/simulative)

The short-term ageing of bitumen, which is generally taken to be the ageing during mixing, transport, laying and compaction of the asphalt made with that bitumen, is assessed using the rolling thin film oven test (RTFOT). RTFOT, standardised in Europe as EN 12607-1 (CEN, 2014c) and in America as ASTM D2872 (ASTM, 2012a), involves a film of bitumen being applied to the inside of bottles which are rotated in an oven at an elevated temperature for a given time with a constant supply of air. There are alternatives to RTFOT in the thin film oven test (TFOT) to EN 12607-2 (CEN, 2014d) and the rotating flask test (RFT) to EN 12607-3 (CEN, 2014e).

The long-term ageing of bitumen, which is generally taken to be the ageing during the service life of the asphalt made with that bitumen, is assessed using pressure ageing vessel (PAV) conditioning. PAV, standardised in Europe as EN 14769 (CEN, 2012q) and in America as ASTM D6521

(ASTM, 2013b), involves ageing trays of bitumen at elevated temperatures under pressurised conditions in a PAV.

The effects of both types of ageing are assessed in terms of the proportional change in mass or of the change in the binder's penetration to EN 1426 (CEN, 2015c), softening point to EN 1427 (CEN, 2015d) or dynamic viscosity.

7.5.3 SATS conditioning (simulative)

The SATS test for asphalt, described in Section 7.2.5, is similar to the PAV test for binder and can be regarded as an ageing conditioning procedure. However, the test as currently drafted is limited to a specific type of mixture.

7.5.4 Asphalt ageing (simulative)

A draft set of procedures is being developed by CEN for conditioning to address oxidative ageing that, when published, will be Part 52 of the EN 12697 series, although probably as a technical specification (draft for development) rather than a full test standard. The draft includes procedures for conditioning either the loose mixture or the compacted mixture. Conditioning the loose mixture allows the conditioning to be more uniform across the whole sample but making samples from aged loose mixtures becomes more difficult whilst conditioning compacted samples means that there are compacted samples that can be tested but may age the exterior more than the centre to a greater extent than would occur in practice.

To condition the loose asphalt, the mixture is placed into a pan and conditioned within a heating cabinet with forced air ventilation for a specific duration at elevated temperature to accelerate ageing due to oxidation to address the short-term ageing potential. In addition, pressure can be applied for further acceleration of conditioning to address the long-term ageing potential.

To condition compacted specimens, either specimens are placed into a pan and conditioned within a heating cabinet with forced air ventilation for a specific duration at a specific temperature, or specimens are placed in a triaxial cell and a forced flow of ozone-enriched compressed air through the specimen is used to condition the specimen for a specific duration at a specific temperature.

7.5.5 Specifying against ageing

There is no way of specifying the resistance of asphalt to all forms of ageing that is generally accepted as being reliable. The short- and long-term ageing of bitumen by the RTFOT and PAV tests are accepted as reliable and are useful guides of the ageing of that component of the asphalt but the extents of those ageings are not generally that variable as to require specification.

It is hoped that the asphalt ageing test will prove as reliable as, and gain the acceptance of, the RTFOT and PAV tests in time after the test is published.

7.6 INTERLAYER BOND

7.6.1 Issue being addressed

The lack of interlayer bond can impair the serviceability of the asphalt pavement because each layer will act separately rather than as a single unit (Nicholls et al., 2008). Furthermore, any moisture trapped between layers will be available to attack the aggregate/binder bond if there is any sensitivity to moisture. The requirement to ensure adequate attention in asphalt pavements is getting increasing attention, but past experience without anything to assist bonding has shown that, with good workmanship, an adequate bond can be achieved with just good cleanliness. A good bond is necessary, but a bond can be achieved without a tack or bond coat.

Interlayer bond is not a property of a mixture because it will depend on not just the mixture being laid but also on the material onto which it is being laid (which may not necessarily be another asphalt), any tack or bond coat applied between the layers and the workmanship in the laying and compacting of each layer. As such, the interlayer bond is not a property that an asphalt mixture can be CE Marked against in the European system.

A further issue with interlayer bond is that any measurements will depend on the type of force used to break the bond. At the extremes, two very rough surfaces placed on top of each other will exhibit considerable bond if the top surface is pushed or twisted relative to the other, but will have no resistance other than self-weight to being lifted off. Alternatively, two glass panes with water between them will be very difficult to pull straight up but will be able to be easily pushed or rotated relative to each other.

Therefore, when measuring interlayer bond, the substrate, tack or bond coat and the type of force need to be defined before any judgement is made. Nevertheless, CEN is preparing a standard covering a torque bond, a shear bond test and a tensile adhesion test that, when published, will be Part 48 of the EN 12697 series. The method is to be for the bond strength between an asphalt layer and other newly constructed construction layers or existing substrates in road or airfield pavements.

7.6.2 Conventional measures

The main issues that are often used are to require the application of a tack or bond coat before overlaying and cleaning the substrate of detritus before applying the tack/bond coat and again before overlaying. The UK Specification for Highway Works (HA et al., 2008a) requires that 'the surface of all bituminous material shall be kept clean and uncontaminated' and that 'prior to placing bituminous material on any new or existing bound substrate, a bond coat or tack coat shall be applied'. Tack and bond coats are similar in intention, but tack coats generally contain unmodified bitumen while bond coats contain polymer-modified bitumen. The need for tack or bond coats often depends on the condition and age of the substrate, with the need often being omitted when the substrate is freshly laid asphalt.

7.6.3 Torque bond (simulative)

The torque bond test can be undertaken either *in situ* or on cores in the laboratory, allowing the temperature to be controlled. A steel plate is glued to the asphalt surface and rotated with the torque moment being measured. For thick layers, a cylindrical groove can be cut through the upper layer down into the substrate to reduce the torque needed. The test can be carried out immediately after laying the upper layer.

The failure can incur in the surface course material, the substrate material or at the bond, whichever is weaker, as shown diagrammatically in Figure 7.2. The failure plane not necessarily being through the interface applies to the other bond tests as well.

A higher torque value implies a stronger bond. Because bond is not an asphalt property, there are no categories for it in EN 13108. Furthermore, the pre-publication draft of the European test gives no precision data.

7.6.4 Shear bond (simulative)

The shear bond test is a laboratory test in which cylindrical specimens are subjected to direct shear force at a constant shear rate and at a fixed temperature. The development of shear deformation and force are recorded

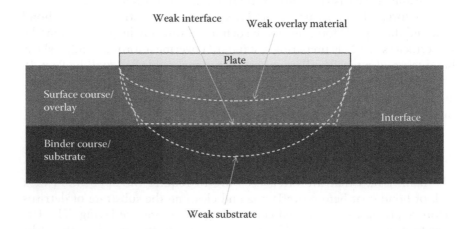

Figure 7.2 Possible failure planes for shear bond test.

with the maximum recorded shear stress being defined as the shear strength at the interface between layers. For thinner layers, a grooved metal plate extension can be affixed to the specimen to minimise bulging in the top layer.

A higher shear stress value implies a stronger bond. Because bond is not an asphalt property, there are no categories for it in EN 13108. The only precision data given in the pre-publication draft of the European test are

- A repeatability standard deviation of 0.074 plus 0.04 times the level for the shear bond strength
- A reproducibility standard deviation of 0.037 plus 0.11 times the level for the shear bond strength
- A repeatability standard deviation of 0.38 mm for the displacement at peak shear stress
- A reproducibility standard deviation of 0.76 mm for the displacement at peak shear stress

7.6.5 Tensile adhesion (simulative)

The tensile adhesion test, which determines the adhesion between the surface layer and substrate perpendicular to the plane of the specimen, involves gluing a plunger to the surface of the top layer and pulling it off at a constant strain rate and fixed temperature. The adhesive tension strength is defined as the maximum force divided by the tension area.

A higher tension strength value implies a stronger bond. Because bond is not an asphalt property, there are no categories for it in EN 13108. Furthermore, the pre-publication draft of the European test gives no precision data.

7.6.6 Specifying for interlayer bond

The specification of interlayer bond is difficult because measuring the potential in the laboratory will be expensive because it requires a combination of substrate, bonding medium (if any) and new overlay that is likely to be practically unique to each site while measurement of the bond *in situ* is destructive. In general, conventional measures are all that is required. When the bond does need to be determined or checked, any value of bond is adequate if the failure is not along the interface because then bond is not the weak link in the chain.

The tensile adhesive bond becomes less important as the layer of asphalt being laid becomes thicker because the mass of that material will weight itself down, even without any tensile adhesion. Therefore, the tensile adhesion test is more applicable for thinner surface course layers.

Both the torque bond and shear bond test measure much the same aspect of bond and can be used to assess the bond required of thicker layers as well as thinner ones. However, the toque bond test has the advantage that it can be used as an *in situ* test.

REFERENCES

American Association of State Highway and Transportation Officials. 1984. Standard method of test for coating and stripping of bitumen-aggregate mixtures. *AASHTO T-182*. Washington, DC: AASHTO.

ASTM International. 2012a. Standard test method for effect of heat and air on a moving film of asphalt (rolling thin-film oven test). *ASTM D2872-12e1*. West Conshohocken, PA: ASTM International.

ASTM International. 2012b. Standard practice for effect of water on bituminous-coated aggregate using boiling water. *D3625/D3625M-12*. West Conshohocken, PA: ASTM International.

ASTM International. 2013b. Standard practice for accelerated aging of asphalt binder using a pressurized aging vessel (PAV). *ASTM D6521-13*. West Conshohocken, PA: ASTM International.

Comité Européen de Normalisation. 2005. Bituminous mixtures – Test methods for hot mix asphalt – Part 43: Resistance to fuel. *EN 12697-43:2005*. London: BSI; Berlin: DIN; Paris: AFNOR; and other European standards institutions.

Comité Européen de Normalisation. 2008c. Bituminous mixtures – Test methods for hot mix asphalt – Part 12: Determination of water sensitivity of bituminous specimens. *EN 12697-12:2008*. London: BSI; Berlin: DIN; Paris: AFNOR; and other European standards institutions.

Comité Européen de Normalisation. 2012g. Bituminous mixtures – Test methods for hot mix asphalt – Part 40: In situ drainability. *EN 12697-40:2012*. London: BSI; Berlin: DIN; Paris: AFNOR; and other European standards institutions.

Comité Européen de Normalisation. 2012n. Bituminous mixtures – Test methods for hot mix asphalt – Part 11: Determination of the affinity between aggregate and bitumen. *EN 12697-11:2012*. London: BSI; Berlin: DIN; Paris: AFNOR; and other European standards institutions.

Comité Européen de Normalisation. 2012o. Bituminous mixtures – Test methods for hot mix asphalt – Part 45: Saturation Ageing Tensile Stiffness (SATS) conditioning test. *EN 12697-45:2012*. London: BSI; Berlin: DIN; Paris: AFNOR; and other European standards institutions.

Comité Européen de Normalisation. 2012p. Bituminous mixtures – Test methods for hot mix asphalt – Part 19: Permeability of specimen. *EN 12697-19:2012*. London: BSI; Berlin: DIN; Paris: AFNOR; and other European standards institutions.

Comité Européen de Normalisation. 2012q. Bitumen and bituminous binders – Accelerated long-term ageing conditioning by a pressure ageing vessel (PAV). *EN 14769:2012*. London: BSI; Berlin: DIN; Paris: AFNOR; and other European standards institutions.

Comité Européen de Normalisation. 2013b. Bituminous mixtures – Test methods for hot mix asphalt – Part 41: Resistance to de-icing fluids. *EN 12697-41:2013*. London: BSI; Berlin: DIN; Paris: AFNOR; and other European standards institutions.

Comité Européen de Normalisation. 2014c. Bitumen and bituminous binders – Determination of the resistance to hardening under the influence of heat and air – Part 1: RTFOT method. *EN 12607-1:2014*. London: BSI; Berlin: DIN; Paris: AFNOR; and other European standards institutions.

Comité Européen de Normalisation. 2014d. Bitumen and bituminous binders – Determination of the resistance to hardening under the influence of heat and air – Part 2: TFOT method. *EN 12607-2:2014*. London: BSI; Berlin: DIN; Paris: AFNOR; and other European standards institutions.

Comité Européen de Normalisation. 2014e. Bitumen and bituminous binders – Determination of the resistance to hardening under the influence of heat and air – Part 3: RFT method. *EN 12607-3:2014*. London: BSI; Berlin: DIN; Paris: AFNOR; and other European standards institutions.

Comité Européen de Normalisation. 2015c. Bitumen and bituminous binders – Determination of softening point – Ring and Ball method. *EN 1427:2015*. London: BSI; Berlin: DIN; Paris: AFNOR; and other European standards institutions.

Comité Européen de Normalisation. 2015d. Bituminous mixtures – Test methods for hot mix asphalt – Part 2: Determination of particle size distribution. *EN 12697-2:2015*. London: BSI; Berlin: DIN; Paris: AFNOR; and other European standards institutions.

Comité Européen de Normalisation. 2016a. Bituminous mixtures – Material specifications – Part 1: Asphalt concrete. *EN 13108-1:2016*. London: BSI; Berlin: DIN; Paris: AFNOR; and other European standards institutions.

Comité Européen de Normalisation. 2016b. Bituminous mixtures – Material specifications – Part 7: Porous asphalt. *EN 13108-7:2016*. London: BSI; Berlin: DIN; Paris: AFNOR; and other European standards institutions.

Daines, M E. 1995. Tests for voids and compaction in rolled asphalt surfacings. *TRL Project Report PR78*. Wokingham: TRL Limited.

Grenfell, J R A, G D Airey, A C Collop, R C Elliott and J C Nicholls. 2011. Assessment of asphalt durability tests; Part 1: Widening the applicability of the SATS test. *TRL Published Project Report PPR535*. Wokingham: TRL Limited.

Hunter, R N, A Self and J Read. 2015. *The Shell Bitumen Handbook*. 6th edition. London: ICE Publishing.

Kandhal, P S, C W Lubold and F L Roberts. 1989. Water damage to asphalt overlays: Case histories. In *Proceedings of the Association of Asphalt Paving Technologists*, Vol. 58, 1989, pp. 40–67. St Paul, MN: AAPT.

Kringos, N, H Azari and A Scarpas. 2009. Identification of parameters related to moisture conditioning that cause variability in modified Lottman test. *Transportation Research Record: Journal of the Transportation Research Board, No. 2127*, pp.1–11. Washington DC: TRB.

Nicholls, J C. 1997. Review of UK porous asphalt trials. *TRL Report TRL264*. Wokingham: TRL Limited.

Nicholls, J C and I G Carswell. 2001. Effectiveness of edge drainage details for use with porous asphalt. *TRL Report TRL376*. Wokingham: TRL Limited.

Nicholls, J C, J Harper, K Green and R C Elliott. 2011. Assessment of asphalt durability tests; Part 3, Review of SATS test to evaluate existing base layers. *TRL Published Project Report PPR537*. Wokingham: TRL Limited.

Nicholls, J C, M J McHale and R D Griffiths. 2008. Best practice guide for durability of asphalt pavements. *TRL Road Note RN42*. Wokingham: TRL Limited.

The Highways Agency, Transport Scotland, Welsh Assembly Government and The Department for Regional Development, Northern Ireland. 2008a. Specification for Highway Works, Series 900, Road Pavements – Bituminous Bound Materials. Manual of Contract Documents for Highway Works, Volume 1. London: The Stationery Office. www.standardsforhighways. co.uk/ha/standards/mchw/vol1/pdfs/series_0900.pdf

Varveri, A, S Avgerinopoulos, T Scarpas, C Nicholls, K Mollenhauer, C McNally, A Gibney and A Tabaković. 2014. Laboratory study on moisture and ageing susceptibility characteristics of RA and WMA mixtures. *EARN deliverable D7*. www.trl.co.uk/solutions/road-rail-infrastructure/sustainable-infrastructure/earn/

Chapter 8

Sustainability

8.1 OVERVIEW

The term 'sustainability' can be used to mean a number of things and covers issues in addition to minimising both carbon dioxide production and the use of virgin resources. However, the issues are generally not covered explicitly in specifications and, when they are, the requirements involve recipe-type restrictions rather than the measurement of a property. Nevertheless, within a world that is becoming more conscious of the limitations of most resources and the damage that man has been doing to the environment, the issues do need to be considered in specifications. Therefore, several that should be considered are discussed briefly here.

Asphalt is a remarkable material that can be durable, is 100% recyclable and can incorporate secondary materials (Nicholls et al., 2010). However, it can be made more durable, more of it can be recycled and more appropriate secondary materials can be incorporated. The result will improve its sustainability by reducing its carbon footprint and enhancing other life-cycle benefits. However, care and attention is needed in the design, construction and maintenance of the material if these benefits are to be fully realised.

8.2 DURABILITY

Durability is a term that everybody thinks they understand but for which everyone has a different definition. Definitions, which differentiate between asphalt durability and asphalt pavement durability, are put forward in TRL Road Note 42 (Nicholls et al., 2008) as

Asphalt Durability
Maintenance of the structural integrity of compacted material over its
*   expected service-life when exposed to the effects of the environment*
*   (water, oxygen, sunlight) and traffic loading*

Pavement Durability
Retention of a satisfactory level of performance over the structure's
  expected service-life without major maintenance for all properties
  that are required for the particular road situation in addition to
  asphalt durability

The durability of the pavement is the most important aspect for mini-mising both the environmental impact and the traffic disruption caused by subsequent maintenance. It is closely aligned to the serviceability properties discussed in Chapter 7 but not limited to them. The use of a durable asphalt mixture should delay the need for maintenance and/or replacement and, hence, often reduce the emissions and need for virgin material in the life of a pavement as much or more than direct actions to reduce these issues when mixing and laying a mixture.

The main means of achieving good durability are to apply the appropri-ate properties as discussed in previous chapters with due consideration for any ageing or possible change of situation. However, additional aspects that need to be considered to achieve good durability in addition to such standard properties are (Nicholls et al., 2010):

- Suitable foundations to the pavement or foundations made suitable at minimum environmental cost using such techniques as cracking and seating on existing concrete pavements
- Drainage capable of keeping water away from the asphalt as far as practicable
- Asphalt that is fully compacted

The expected durability of pavements, particularly surface courses as the most exposed part of the pavement, is relatively limited but could eas-ily be enhanced by simple measures. These measures include the wider use of bond coats and ensuring that there is an adequate binder film thickness. However, bitumen needs to be about 8 mm thick for it to provide a water-proofing layer, a thickness that is not practical because it would introduce other problems (Hunter et al., 2015). The use of binder modifiers is gener-ally designed to enhance certain physical properties, but their effect on the durability of the resulting asphalt and the environmental cost of their production need to be assessed. However, any procedure to enhance dura-bility may have to be moderated because of the conflicting requirements for the performance of other required properties.

8.3 RECYCLING

Recycling should be the easiest method of improving sustainability (Nicholls et al., 2010). Although 100% of reclaimed asphalt (RA) can be recycled, mixtures are not currently produced with 100% RA (which would require very careful selection in order to have the required aggregate grading, the required binder content and binder of the required grade). Nevertheless, the proportions being incorporated are increasing steadily, even in the surface course layer. The open-loop use of RA from layers lower in the pavement than the layer the RA came from has become routine, but higher-level use back into the same layer, known as closed-loop reuse, needs to be further encouraged. One particular advantage with recycling is the reduction of transport to move the component materials to where they are needed.

Guidance is available for recycling, even into surface course mixtures (Carswell et al., 2010). The main difference is that additional testing may be needed to ensure the consistency of the RA. The RA will be more consistent if it is taken from a single source that has a consistent mixture or from many different sources that have been well mixed, the latter being restricted to surface course sources if it is to be used in the surface course.

Any RA that is to be used needs to exclude tar (unless going into a cold asphalt mixture) and asbestos fibres, both of which were used in pavements prior to being discovered to be carcinogenic. RA also needs to have a minimal amount of foreign material, as standardised in Europe as EN 12697-42 (CEN, 2012r), where foreign matter means anything not usually used in the manufacture of asphalt. Polymer-modified binders are not considered foreign matter because they are used in asphalt and it has been found that different modifiers in terms of a different modifier in the new mixture to that in the RA can be mixed successfully (Carswell et al., 2005).

However, as well as considering what recycling is permitted or required for the current asphalt mixture, there is also the consideration of the suitability of that mixture for future recycling. Generally, the asphalt for major works tends to be acceptable but often repairs and reinstatements do impair the recyclability of a pavement when the repair material is different from the original. The worst offenders are the cement- and resin-based mixtures that are marketed and used for pothole repairs.

8.4 SECONDARY AGGREGATES

The use of secondary aggregates is an alternative to using RA in strategies to reduce the extraction of virgin aggregate. The use of secondary materials for filler tends to be counterproductive in that there tends to be excess production of suitable quarry fines resulting from crushing aggregate into the coarse particle sizes needed (Nicholls et al., 2010). However, some use of very fine material, an example being cement kiln dust, may have other environmental benefits (Nicholls et al., 2006b). The incorporation of secondary material for the coarser fractions has greater potential, with a variety of slags having been successfully used for many years whilst others, like crushed glass, have started to be incorporated more recently (Nicholls and Lay, 2002). One aspect of using secondary materials that needs to be considered for true sustainability is their recyclability, including that of binder modifiers. Asphalt is a rare product in that it is 100% recyclable, and any impairment in that situation would be environmentally damaging. Specifications can ban, permit or require the use of specific or all secondary aggregates.

8.5 CARBON EMISSIONS

There is increasing interest in monitoring and reducing the carbon emissions in pavement construction as well as other aspects of modern life. There are several models that have been developed to do this including the Asphalt Pavement Embodied Carbon Tool (asPECT) which can be downloaded free from the website www.sustainabilityofhighways.org.uk/. When comparing different options, the same model and assumptions need to be used, in particular whether it is from cradle to site, cradle to grave or something else. Cradle to site implies winning, processing, transporting, mixing and laying the components, as shown in Figure 8.1 (Schiavi et al., 2007), whilst cradle to grave also includes future maintenance and disposal or reuse of the asphalt as the pavement ages.

Not all the branches will be used in Figure 8.1 for any specific scheme because polymer-modified binders, RA and secondary aggregates are not always used.

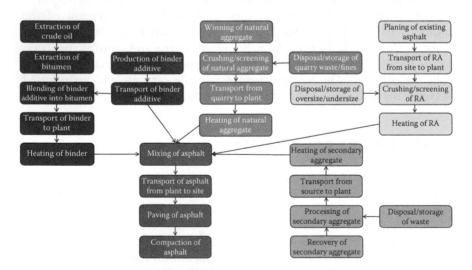

Figure 8.1 Processes to include in assessment of carbon emissions from cradle to site. (After Schiavi, I, I Carswell and M Wayman. 2007. Recycled asphalt in surfacing materials: A case study of carbon dioxide emission savings. *TRL Published Project Report PPR304*. Wokingham: TRL.)

Table 8.1 Summary of lower temperature asphalt systems

Product	Company	Description	Dosage of additive	Country used	Production temperature or reduction ranges	Website
Zeolite additives						
Advera	PQ Corporation	Water containing using Zeolite	0.25% of mixture by mass	USA	(10–30)°C	www.pqcorp.com/products/AdveraWMA.asp
Aspha-Min	Eurovia and MHI	Water containing Zeolite	0.3% of mixture by mass	Worldwide, including France, Germany and USA	(20–30)°C	www.eurovia.fr/en/produit/135.aspx?print=y
Organic (wax) additives						
Asphaltan A Romonta N	Romonta GmbH	Montan wax for mastic asphalt	(1.5–2.0) % of bitumen by mass	Germany	20°C	www.romonta.de/ie4/english/romonta/i_wachse.htm
Asphaltan B		Refined Montan wax with fatty acid amide for rolled asphalt	(2–4) % by mixture by mass	Germany	(20–30)°C	
Sasobit	Sasol	Fischer–Tropsch wax	(2.5–3.0) % of bitumen by mass in Germany; (1.0–1.5) % of bitumen by mass in USA	Worldwide, including EU, RSA and USA	(20–30)°C	www.sasolwax.us.com/sasobit.html

(Continued)

Table 8.1 (Continued) Summary of lower temperature asphalt systems

Product	Company	Description	Dosage of additive	Country used	Production temperature or reduction ranges	Website
Sasolwax Flex		Fischer–Tropsch wax plus polymer (choice of type)	Unspecified		At least 28°C	www.sasolwax.com/More_about_Sasolwax_Flex.html
Fatty acid derivative additives						
Hypertherm	Coco Asphalt Engineering	Fatty acid derivative	Unspecified	Canada	Unspecified	www.cocoasphaltengineering.com/warm_mix.aspx
Licomont BS 100	Clariant	Fatty acid amide wax	3% of bitumen by mass	Germany	(20–30)°C	http://www.clariant.com/en/Solutions/Products/2014/03/18/16/33/Licomont-BS-100-granules
Chemical additives						
Cecabase RT	CECA Arkema group	Chemical package	(0.2–0.4)% of mixture by mass	France and USA	120°C	http://www.cecachemicals.com/en/products/ceca-product-finder/range-page/Cecabase-RT/
Ecoflex or 3ELT	Colas	Unspecified additive	Unspecified	France	(30–40)°C	

(Continued)

Table 8.1 (Continued) Summary of lower temperature asphalt systems

Product	Company	Description	Dosage of additive	Country used	Production temperature or reduction ranges	Website
Evotherm DAT	MeadWestvaco	Chemical package plus water	30% of binder by mass	Worldwide, including France and USA	(85–115)°C	www.meadwestvaco.com/ Products/MWV002106
Evotherm 3G or REVIX		Water-free chemical package	Unspecified	USA	(15–27)°C	
Qualitherm	QPR ShopWorx	Unspecified additive	Unspecified	Canada and USA	Unspecified	
Rediset WMX	Akzo Nobel	Cationic surfactants and organic additive	(1.5–2) % of bitumen by mass	Norway and USA	≥30°C 126°C	www.surfactants.akzonobel. com/asphalt/pdf/Rediset%20 Brochure_0907.PDF
Sübit VR	GKG Mineraloel Handel	Unspecified additive	Unspecified	Germany	Unspecified	www.gkg-oel.de/fileadmin/ gkg-oel/Dokumente/ Produktbeschreibung.pdf
Other specified additives						
Thipoave	Shell	Sulphur plus compaction aid	(30–50) % of bitumen by mass	Worldwide	20°C	
TLA-X	Lake Asphalt of Trinidad and Tobago	Trinidad Lake Asphalt plus modifiers	Unspecified	Worldwide	Unspecified	www.trinidadlakeasphalt.com/ home/products/tla-x-warm- mix-technology.html

(Continued)

Table 8.1 (Continued) Summary of lower temperature asphalt systems

Product	Company	Description	Dosage of additive	Country used	Production temperature or reduction ranges	Website
Emulsions						
ECOMAC	SCREG	Cold mix warmed before laying	Unknown type or quantity	France	c. 45°C	
Evotherm ET	Mead-Westvaco	Chemical bitumen emulsion	Delivered in form of bitumen emulsion	Worldwide, including France and USA	(85–115)°C	www.meadwestvaco.com/ Products/MWV002106
Foaming technology						
Accu-Shear Dual Warm Mix Additive System	Stansteel	Water-based foaming process	Unnecessary	USA	Unspecified	www.stansteel.com/sip.html
Adesco/Madsen Static Inline Vortex Mixer	Adesco/Madsen	Water-based foaming process	Unnecessary	USA	Unspecified	
Aquablack WMA	MAXAM equipment	Water-based foaming process	Unnecessary	USA	Unspecified	http://maxamequipment.com/ AQUABlackWMA.htm

(Continued)

Table 8.1 (Continued) Summary of lower temperature asphalt systems

Product	Company	Description	Dosage of additive	Country used	Production temperature or reduction ranges	Website
AquaFoam	Reliable Asphalt Products	Water-based foaming process	Unnecessary	USA	Unspecified	www.reliableasphalt.com/Default.asp
Double Barrel Green	Astec	Water-based foaming process	Optional anti-stripping agent	USA	(116–135)°C	www.astecinc.com/index.php?option=com_content&view=article&id=117&Itemid=188
ECO-Foam II	Aesco/Madsen	Water-based foaming process	Unnecessary	USA	Unspecified	
NA Foamtec	Foamtec International	Water-based foaming process	(1.5–3.0)% by mass of binder	RSA and USA	Unspecified	
H Grant Warm Mix System	Herman Grant Company	Water-based foaming process	Unnecessary	USA	Unspecified	www.hermangrant.com/warm-mix.htm
LEA (Low Energy Asphalt)	LEACO and McConnaughay	Water-based hot coarse aggregate mixed with wet sand	±0.5% of bitumen by mass of coating and adhesion additive	France, Italy, Spain and USA	≤100°C	–

(Continued)

Table 8.1 (Continued) Summary of lower temperature asphalt systems

Product	Company	Description	Dosage of additive	Country used	Production temperature or reduction ranges	Website
LEAB	BAM Wegen bv	Water-based mixing of aggregates below water boiling point	0.1% of bitumen by mass of coating and adhesion additive	Netherlands	90°C	www.bamwegen.nl/sites/www. bamwegen.nl/files/site_images/ LEAB%20-%20Asphalt%20 English.pdf
LT Asphalt	Nynas	Water-based binder foaming with hygrophilic filler	Hygroscopic filler at (0.5–1.0) % of mixture by mass	Worldwide, including Italy and Netherlands	90°C	http://nyport.nynas.com/ Apps/1112.nsf/wpis/GB_EN_ LT-Asphalt/$File/LT-Asphalt_ GB_EN_PIS.pdf
Meeker Warm Mix Asphalt System	Meeker Equipment	Water-based foaming process	Unnecessary	Spain	Unspecified	www.meekerequipment.com/ new_warmmixad1.html
Ultrafoam GX	Gencor Industries	Water-based foaming process	Unnecessary	USA	Unspecified	http://gencorgreenmachine.com
Warm Mix Asphalt System	Terex Roadbuilding	Water-based foaming process	Unnecessary	USA	<32°C	www.terexrb.com/default. aspx?pgID=308

(Continued)

Table 8.1 (Continued) Summary of lower temperature asphalt systems

Product	Company	Description	Dosage of additive	Country used	Production temperature or reduction ranges	Website
Other processes						
Low Emission Asphalt	McConnaughay Technologies	Combination of chemical and water-based foaming technology	0.4% of bitumen by mass	USA	90°C	
WAM-Foam	Shell, Kolo Veidekke	Foaming process using two binder grades	Anti-stripping agents can be added to soften binder	Norway, France, Canada, Italy, Luxemburg, Netherlands, Sweden, Switzerland, UK and USA	(110– 120)°C	

Source: Nicholls, J C, S Cassidy, C McNally, K Mollenhauer, R Shahmohammadi, A Tabaković, R Taylor, A Varveri and M Wayman. 2014. Final report on effects of using reclaimed asphalt and/or lower temperature asphalt on the road network. EARN Project for CEDR, Deliverable No. D9. www.trl.co.uk/solutions/road-rail-infra-structure/sustainable-infrastructure/earn/

A maximum limit on the carbon emissions can be used in specifications or a ranking of those emissions used to compare different options.

8.6 LOWER TEMPERATURE ASPHALT

Until relatively recently, hot mix asphalt was the norm with lower temperature options being regarded as inferior because of the perceived need for high-temperature mixing to achieve complete coverage of the aggregate particles in the mixing and adequate compaction on site. With modern developments, these issues have been addressed and there are a multitude of different systems that can produce asphalt successfully at lower temperatures. Some typical systems that are available in Europe and/or America, which are generally proprietary, are listed in Table 8.1 (Nicholls et al., 2014). However, many engineers still consider that these systems do not have all the properties of traditional hot mix asphalt, which is correct for various properties with some systems.

Specifications can bar, permit or encourage the use of some or all of such lower temperature asphalt systems. Because of the variety of systems, it is difficult to produce a general specification covering all aspects of all, although some have been produced (Nicholls et al., 2013). In general, extra care needs to be taken to ensure that aggregate particles are fully covered with binder, that adequate compaction is achieved and that any water added as part of the system does not affect the durability of the asphalt.

REFERENCES

Carswell, I, J C Nicholls, R C Elliott, J Harris and D Strickland. 2005. Feasibility of recycling thin surfacing back into thin surfacing systems. *TRL Report TRL645*. Wokingham: TRL Limited.

Carswell, I, J C Nicholls, I Widyatmoko, J Harris and R Taylor. 2010. Best practice guide for recycling into surface course. *TRL Road Note RN43*. Wokingham: TRL Limited.

Comité Européen de Normalisation. 2012r. Bituminous mixtures – Test methods for hot mix asphalt – Part 42: Amount of foreign matter in reclaimed asphalt. *EN 12697-42:2012*. London: BSI; Berlin: DIN; Paris: AFNOR; and other European standards institutions.

Hunter, R N, A Self and J Read. 2015. *The Shell Bitumen Handbook*. 6th edition. London: ICE Publishing.

Nicholls, J C, H K Bailey, N Ghazireh and D H Day. 2013. Specification for low temperature asphalt mixtures. *TRL Published Project Report PPR666*. Wokingham: TRL Limited.

Nicholls, J C, I Carswell, M Wayman and J M Reid. 2010. Increasing the environmental sustainability of asphalt. *TRL Insight Report INS007*. Wokingham: TRL Limited.

Nicholls, J C, S Cassidy, C McNally, K Mollenhauer, R Shahmohammadi, A Tabaković, R Taylor, A Varveri and M Wayman. 2014. Final report on effects of using reclaimed asphalt and/or lower temperature asphalt on the road network. *EARN Project for CEDR, Deliverable No. D9.* www.trl.co.uk/ solutions/road-rail-infrastructure/sustainable-infrastructure/earn/

Nicholls, J C and J Lay. 2002. Crushed glass in asphalt for binder course and roadbase layers. In *3rd International Conference Bituminous Mixtures and Pavements.* Thessaloniki, Greece: Aristotle University of Thessaloniki.

Nicholls, J C, M J McHale and R D Griffiths. 2008. Best practice guide for durability of asphalt pavements. *TRL Road Note RN42.* Wokingham: TRL Limited.

Nicholls, J C, J M Reid, C D Whiteoak and M Wayman. 2006b. Cement kiln dust (CKD) as filler in asphalt. *TRL Report TRL659.* Wokingham: TRL Limited.

Schiavi, I, I Carswell and M Wayman. 2007. Recycled asphalt in surfacing materials: A case study of carbon dioxide emission savings. *TRL Published Project Report PPR304.* Wokingham: TRL.

Chapter 9

Summary

Asphalt is a remarkable and useful material for the construction of highway, airfield and other pavements. The myriad of properties that can be provided is wide ranging with the extent that each aspect is required depending on the location and geometry of the site, the traffic expected on the pavement, the layer within the pavement and the design concept for the pavement. The particular design procedure consideration that is relevant is whether it is for conventional dense pavement or for a pervious pavement in a sustainable drainage system (SUDS).

Overall, this review of the properties of asphalt demonstrates that it is possible to obtain almost any desirable property in a particular mixture. However, the properties required of asphalt are not always the same as the properties that can be measured. When specifying or describing properties like durability or workability, those conceptual properties need to be converted into more distinct and measurable surrogate properties. Properties that can be used to define the properties of an asphalt can sometimes be fundamental but often have to be simulative, surrogate or even conventional.

When deciding on a specification for an asphalt, there are three simple questions that need to be asked:

1. What properties are required for the situation (traffic, geometry, layer, service life, etc.) for which the asphalt is intended?
2. For each property, what level of performance (if any) is justifiable?
3. For each property to be specified, how is compliance going to be checked?

Ultimately, nothing should be called for that is not necessary or for which there will be no check that it has been supplied.

However, no asphalt can have the highest level for every potential property, some of the properties being effectively mutually exclusive. Furthermore, every performance requirement is likely to reduce the options for the supplier and increase the cost of the material. Therefore, it should be the aim of engineers specifying or supplying asphalts to specify or define all the necessary properties at the required level and not to specify unnecessary properties or properties at unnecessary levels just 'to be sure'.

If an engineer is someone who can do for 10 pence what any fool can do for a pound, then a highway engineer is someone who provides all the properties required at the appropriate level without providing unnecessary properties and levels.

Index

Printed and bound by CPI Group (UK) Ltd, Croydon, CR0 4YY
01/11/2024
01782621-0019